T0324942

Graduate Texts in Mathematics 89

Editorial Board
J.H. Ewing F.W. Gehring P.R. Halmos

BOOKS OF RELATED INTEREST BY SERGE LANG

Linear Algebra, Third Edition
1987, ISBN 96412-6

Undergraduate Algebra, Second Edition
1990, ISBN 97279-X

Complex Analysis, Third Edition
1993, ISBN 97886-0

Real and Functional Analysis, Third Edition
1993, ISBN 94001-4

Algebraic Number Theory, Second Edition
1994, ISBN 94225-4

Introduction to Complex Hyperbolic Spaces
1987, ISBN 96447-9

OTHER BOOKS BY LANG PUBLISHED BY
SPRINGER-VERLAG

Introduction to Arakelov Theory • Riemann-Roch Algebra (with William Fulton) • Complex Multiplication • Introduction to Modular Forms • Modular Units (with Daniel Kubert) • Fundamentals of Diophantine Geometry • Elliptic Functions • Number Theory III • Cyclotomic Fields I and II • $SL_2(R)$ • Abelian Varieties • Differential and Riemannian Manifolds • Undergraduate Analysis • Elliptic Curves: Diophantine Analysis • Introduction to Linear Algebra • Calculus of Several Variables • First Course in Calculus • Basic Mathematics • Geometry: A High School Course (with Gene Murrow) • Math! Encounters with High School Students • The Beauty of Doing Mathematics • THE FILE

Serge Lang

Introduction to Algebraic
and Abelian Functions

Second Edition

Springer-Verlag

New York Berlin Heidelberg London Paris
Tokyo Hong Kong Barcelona Budapest

Serge Lang
Department of Mathematics
Yale University
New Haven, Connecticut 06520
USA

AMS Classifications: 14HOJ, 14K25

With 9 illustrations.

Library of Congress Cataloging in Publication Data
Lang, Serge, 1927–
    Introduction to algebraic and abelian functions.
    (Graduate texts in mathematics; 89) Bibliography: p. 165 Includes index.
    1. Functions, Algebraic. 2. Functions, Abelian.
I. Title. II. Series. QA341.L32 1982 515.9'83 82-5733 AACR2

The first edition of *Introduction to Algebraic and Abelian Functions* was published
in 1972 by Addison-Wesley Publishing Co., Inc.

Typeset by Interactive Composition Corporation, Pleasant Hill, CA.
Printed and bound by Braun-Brumfield, Ann Arbor, MI.
Printed in the United States of America.

9 8 7 6 5 4 3 2 (Second corrected printing, 1995)

ISBN 0-387-90710-6 Springer-Verlag New York Berlin Heidelberg
ISBN 3-540-90710-6 Springer-Verlag Berlin Heidelberg New York

Introduction

This short book gives an introduction to algebraic and abelian functions, with emphasis on the complex analytic point of view. It could be used for a course or seminar addressed to second year graduate students.

The goal is the same as that of the first edition, although I have made a number of additions. I have used the Weil proof of the Riemann-Roch theorem since it is efficient and acquaints the reader with adeles, which are a very useful tool pervading number theory.

The proof of the Abel-Jacobi theorem is that given by Artin in a seminar in 1948. As far as I know, the very simple proof for the Jacobi inversion theorem is due to him. The Riemann-Roch theorem and the Abel-Jacobi theorem could form a one semester course.

The Riemann relations which come at the end of the treatment of Jacobi's theorem form a bridge with the second part which deals with abelian functions and theta functions. In May 1949, Weil gave a boost to the basic theory of theta functions in a famous Bourbaki seminar talk. I have followed his exposition of a proof of Poincaré that to each divisor on a complex torus there corresponds a theta function on the universal covering space. However, the correspondence between divisors and theta functions is not needed for the linear theory of theta functions and the projective embedding of the torus when there exists a positive non-degenerate Riemann form. Therefore I have given the proof of existence of a theta function corresponding to a divisor only in the last chapter, so that it does not interfere with the self-contained treatment of the linear theory.

The linear theory gives a good introduction to abelian varieties, in an analytic setting. Algebraic treatments become more accessible to the reader who has gone through the easier proofs over the complex numbers. This includes the duality theory with the Picard, or dual, abelian manifold.

I have included enough material to give all the basic analytic facts necessary in the theory of complex multiplication in Shimura-Taniyama, or my more recent book on the subject, and have thus tried to make this topic accessible at a more elementary level, provided the reader is willing to assume some algebraic results.

I have also given the example of the Fermat curve, drawing on some recent results of Rohrlich. This curve is both of intrinsic interest, and gives a typical setting for the general theorems proved in the book. This example illustrates both the theory of periods and the theory of divisor classes. Again this example should make it easier for the reader to read more advanced books and papers listed in the bibliography.

New Haven, Connecticut   SERGE LANG

Contents

Chapter IV
The Theorem of Abel-Jacobi

Chapter V
Periods on the Fermat Curve

Chapter VI
Linear Theory of Theta Functions

Chapter VII
Homomorphisms and Duality

Chapter VIII
Riemann Matrices and Classical Theta Functions

Chapter IX
Involutions and Abelian Manifolds of Quaternion Type

Chapter X
Theta Functions and Divisors

Bibliography

Index

CHAPTER I
The Riemann-Roch Theorem

§1. Lemmas on Valuations

We recall that a **discrete valuation ring** $\mathfrak{o}$ is a principal ideal ring (and there-fore a unique factorization ring) having only one prime. If t is a generator of this prime, we call t a **local parameter**. Every element $x \neq 0$ of such a ring can be expressed as a product

$$x = t^r y,$$

where r is an integer ≥ 0, and y is a unit. An element of the quotient field K has therefore a similar expression, where r may be an arbitrary integer, which is called the **order** or **value** of the element. If $r > 0$, we say that x has a **zero** at the valuation, and if $r < 0$, we say that x has a **pole**. We write

$$r = v_{\mathfrak{o}}(x), \quad \text{or} \quad v(x), \quad \text{or} \quad \text{ord}_{\mathfrak{o}}(x).$$

Let $\mathfrak{p}$ be the maximal ideal of $\mathfrak{o}$. The map of K which is the canonical map $\mathfrak{o} \to \mathfrak{o}/\mathfrak{p}$ on $\mathfrak{o}$, and sends an element $x \notin \mathfrak{o}$ to ∞, is called the **place** of the valuation.

We shall take for granted a few basic facts concerning valuations, all of which can be found in my *Algebra*. Especially, if E is a finite extension of K and $\mathfrak{o}$ is a discrete valuation ring in K with maximal ideal $\mathfrak{p}$, then there exists a discrete valuation ring $\mathfrak{O}$ in E, with prime $\mathfrak{P}$, such that

$$\mathfrak{o} = \mathfrak{O} \cap K \quad \text{and} \quad \mathfrak{p} = \mathfrak{P} \cap K.$$

If u is a prime element of $\mathfrak{O}$, then $t\mathfrak{O} = u^e \mathfrak{O}$, and e is called the **ramifica-**

tion index of $\mathfrak{O}$ over $\mathfrak{o}$ (or of $\mathfrak{P}$ over $\mathfrak{p}$). If $\Gamma_\mathfrak{O}$ and $\Gamma_\mathfrak{o}$ are the value groups of these valuation rings, then $(\Gamma_\mathfrak{O} : \Gamma_\mathfrak{o}) = e$.

We say that the pair $(\mathfrak{O},\mathfrak{P})$ **lies above** $(\mathfrak{o},\mathfrak{p})$, or more briefly that $\mathfrak{P}$ **lies above** $\mathfrak{p}$. We say that $(\mathfrak{O},\mathfrak{P})$ is **unramified above** $(\mathfrak{o},\mathfrak{p})$, or that $\mathfrak{P}$ is **unramified** above $\mathfrak{p}$, if the ramification index is equal to 1, that is $e = 1$.

Example. Let k be a field and t transcendental over k. Let $a \in k$. Let $\mathfrak{o}$ be the set of rational functions

$$f(t)/g(t), \quad \text{with} \quad f(t), g(t) \in k[t] \quad \text{such that} \quad g(a) \neq 0.$$

Then $\mathfrak{o}$ is a discrete valuation ring, whose maximal ideal consists of all such quotients such that $f(a) = 0$. This is a typical situation. In fact, let k be algebraically closed (for simplicity), and consider the extension $k(x)$ obtained with one transcendental element x over k. Let $\mathfrak{o}$ be a discrete valuation ring in $k(x)$ containing k. Changing x to $1/x$ if necessary, we may assume that $x \in \mathfrak{o}$. Then $\mathfrak{p} \cap k[x] \neq 0$, and $\mathfrak{p} \cap k[x]$ is therefore generated by an irreducible polynomial $p(x)$, which must be of degree 1 since we assumed k algebraically closed. Thus $p(x) = x - a$ for some $a \in k$. Then it is clear that the canonical map

$$\mathfrak{o} \to \mathfrak{o}/\mathfrak{p}$$

induces the map

$$f(x) \mapsto f(a)$$

on polynomials, and it is then immediate that $\mathfrak{o}$ consists of all quotients $f(x)/g(x)$ such that $g(a) \neq 0$; in other words, we are back in the situation described at the beginning of the example.

Similarly, let $\mathfrak{o} = k[[t]]$ be the ring of formal power series in one variable. Then $\mathfrak{o}$ is a discrete valuation ring, and its maximal ideal is generated by t. Every element of the quotient field has a formal series expansion

$$x = a_{-m}t^{-m} + \cdots + a_{-1}t^{-1} + a_0 + a_1 t + a_2 t^2 + \cdots,$$

with coefficients $a_i \in k$. The place maps x on the value a_0 if x does not have a pole.

In the applications, we shall study a field K which is a finite extension of a transcendental extension $k(x)$, where k is algebraically closed, and x is transcendental over k. Such a field is called a **function field in one variable**. If that is the case, then the residue class field of any discrete valuation ring $\mathfrak{o}$ containing k is equal to k itself, since we assumed k algebraically closed.

Proposition 1.1. *Let E be a finite extension of K. Let $(\mathfrak{O},\mathfrak{P})$ be a discrete* *valuation ring in E above* $(\mathfrak{o}, \mathfrak{p})$ *in K. Suppose that* $E = K(y)$ *where y is the root of a polynomial* $f(Y) = 0$ *having coefficients in* $\mathfrak{o}$, *leading coefficient 1, such that*

$$f(y) = 0 \quad but \quad f'(y) \not\equiv 0 \bmod \mathfrak{P}.$$

Then $\mathfrak{P}$ *is unramified over* $\mathfrak{p}$.

Proof. There exists a constant $y_0 \in k$ such that $y \equiv y_0 \bmod \mathfrak{P}$. By hypothesis, $f'(y_0) \not\equiv 0 \bmod \mathfrak{P}$. Let $\{y_n\}$ be the sequence defined recursively by

$$y_{n+1} = y_n - f'(y_n)^{-1}f(y_n).$$

Then we leave to the reader the verification that this sequence converges in the completion $K_\mathfrak{p}$ of K, and it is also easy to verify that it converges to the root y since $y \equiv y_0 \bmod \mathfrak{P}$ but y is not congruent to any other root of f and $\mathfrak{P}$. Hence y lies in this completion, so that the completion $E_\mathfrak{P}$ is embedded in $K_\mathfrak{p}$, and therefore $\mathfrak{P}$ is unramified.

We also recall some elementary approximation theorems.

Chinese Remainder Theorem. *Let R be a ring, and let* $\mathfrak{p}_1, \ldots, \mathfrak{p}_n$ *be distinct maximal ideals in that ring. Given positive integers* $r_1, \ldots, r_n$ *and elements* $a_1, \ldots, a_n \in R$, *there exists* $x \in R$ *satisfying the congruences*

$$x \equiv a_i \bmod \mathfrak{p}_i^{r_i} \quad for \; all \; i.$$

For the proof, cf. *Algebra*, Chapter II, §2. This theorem is applied to the integral closure of $k[x]$ in a finite extension.

We shall also deal with similar approximations in a slightly different context, namely a field K and a finite set of discrete valuation rings $\mathfrak{o}_1, \ldots, \mathfrak{o}_n$ of K, as follows.

Proposition 1.2. *If* $\mathfrak{o}_1$ *and* $\mathfrak{o}_2$ *are two discrete valuation rings with quotient field K, such that* $\mathfrak{o}_1 \subset \mathfrak{o}_2$, *then* $\mathfrak{o}_1 = \mathfrak{o}_2$.

Proof. We shall first prove that if $\mathfrak{p}_1$ and $\mathfrak{p}_2$ are their maximal ideals, then $\mathfrak{p}_2 \subset \mathfrak{p}_1$. Let $y \in \mathfrak{p}_2$. If $y \notin \mathfrak{p}_1$, then $1/y \in \mathfrak{o}_1$, whence $1/y \in \mathfrak{o}_2$, a contradiction. Hence $\mathfrak{p}_2 \subset \mathfrak{p}_1$. Every unit of $\mathfrak{o}_1$ is *a fortiori* a unit of $\mathfrak{o}_2$. An element y of $\mathfrak{p}_2$ can be written $y = \pi_1^{\nu_1}u$ where u is a unit of $\mathfrak{o}_1$ and π_1 is an element of order 1 in $\mathfrak{p}_1$. If π_1 is not in $\mathfrak{p}_2$, it is a unit in $\mathfrak{o}_2$, a contradiction. Hence π_1 is in $\mathfrak{p}_2$, and hence so is $\mathfrak{p}_1 = \mathfrak{o}_1\pi_1$. This proves $\mathfrak{p}_2 = \mathfrak{p}_1$. Finally, if u is a unit in $\mathfrak{o}_2$, and is not in $\mathfrak{o}_1$, then $1/u$ is $\mathfrak{p}_1$, and thus cannot be a unit in $\mathfrak{o}_2$. This proves our proposition.

From now on, we assume that *our valuation rings $\mathfrak{o}_i$ ($i = 1, \ldots, n$) are distinct, and hence have no inclusion relations.*

Proposition 1.3. *There exists an element y of K having a zero at $\mathfrak{o}_1$ and a pole at $\mathfrak{o}_j$ ($j = 2, \ldots, n$).*

Proof. This will be proved by induction. Suppose $n = 2$. Since there is no inclusion relation between $\mathfrak{o}_1$ and $\mathfrak{o}_2$, we can find $y \in \mathfrak{o}_2$ and $y \notin \mathfrak{o}_1$. Similarly, we can find $z \in \mathfrak{o}_1$ and $z \notin \mathfrak{o}_2$. Then z/y has a zero at $\mathfrak{o}_1$ and a pole at $\mathfrak{o}_2$ as desired.

Now suppose we have found an element y of K having a zero at $\mathfrak{o}_1$ and a pole at $\mathfrak{o}_2, \ldots, \mathfrak{o}_{n-1}$. Let z be such that z has a zero at $\mathfrak{o}_1$ and a pole at $\mathfrak{o}_n$. Then for sufficiently large r, $y + z^r$ satisfies our requirements, because we have schematically zero plus zero = zero, zero plus pole = pole, and the sum of two elements of K having poles of different order again has a pole.

A high power of the element y of Proposition 1.3 has a high zero at $\mathfrak{o}_1$ and a high pole at $\mathfrak{o}_j$ ($j = 2, \ldots, n$). Adding 1 to this high power, and considering $1/(1 + y^r)$ we get

Corollary. *There exists an element z of K such that $z - 1$ has a high zero at $\mathfrak{o}_1$, and such that z has a high zero at $\mathfrak{o}_j$ ($j = 2, \ldots, n$).*

Denote by ord_i the order of an element of K under the discrete valuation associated with $\mathfrak{o}_i$. We then have the following approximation theorem.

Theorem 1.4. *Given elements $a_1, \ldots, a_n$ of K, and an integer N, there exists an element $y \in K$ such that $\text{ord}_i(y - a_i) > N$.*

Proof. For each i, use the corollary to get z_i close to 1 at $\mathfrak{o}_i$ and close to 0 at $\mathfrak{o}_j$ ($j \neq i$), or rather at the valuations associated with these valuation rings. Then $z_1 a_1 + \cdots + z_n a_n$ has the required property.

In particular, we can find an element y having given orders at the valuations arising from the $\mathfrak{o}_i$. This is used to prove the following inequality.

Corollary. *Let E be a finite algebraic extension of K. Let Γ be the value group of a discrete valuation of K, and Γ_i the value groups of a finite number of inequivalent discrete valuations of E extending that of K. Let e_i be the index of Γ in Γ_i. Then*

$$\sum e_i \leqq [E : K].$$

Proof. Select elements

$$y_{11}, \ldots, y_{1e_1}, \ldots, y_{r1}, \ldots, y_{re_r}$$

of E such that $y_{i\nu}$ ($\nu = 1, \ldots, e_i$) represent distinct cosets of Γ in Γ_i, and have zeroes of high order at the other valuations v_j ($j \neq i$). We contend that the above elements are linearly independent over K. Suppose we have a relation of linear dependence

$$\sum_{i,\nu} c_{i\nu} y_{i\nu} = 0.$$

Say c_{11} has maximal value in Γ, that is, $v(c_{11}) \geqq v(c_{i\nu})$ all i, v. Divide the equation by c_{11}. Then we may assume that $c_{11} = 1$, and that $v(c_{i\nu}) \leqq 1$. Consider the value of our sum taken at v_1. All terms $y_{11}, c_{12} y_{12}, \ldots, c_{1e_1} y_{1e_1}$ have distinct values because the y's represent distinct cosets. Hence

$$v_1(y_{11} + \cdots + c_{1e_1} y_{1e_1}) \geqq v_1(y_{11}).$$

On the other hand, the other terms in our sum have a very small value at v_1 by hypothesis. Hence again by that property, we have a contradiction, which proves the corollary.

§2. The Riemann-Roch Theorem

Let k be an algebraically closed field, and let K be a function field in one variable over k (briefly a **function field**). By this we mean that K is a finite extension of a purely transcendental extension $k(x)$ of k, of transcendence degree 1. We call k the **constant field**. Elements of K are sometimes called **functions**.

By a **prime**, or **point**, of K over k, we shall mean a discrete valuation ring of K containing k (or over k). As we saw in the example of §1, the residue class field of this ring is then k itself. The set of all such discrete valuation rings (i.e., the set of all points of K) will be called a **curve**, whose function field is K. We use the letters P, Q for points of the curve, to suggest geometric terminology.

By a **divisor** (on the curve, or of K over k) we mean an element of the free abelian group generated by the points. Thus a divisor is a formal sum.

$$\mathfrak{a} = \sum n_i P_i = \sum n_P P$$

where P_i are points, and n_i are integers, all but a finite number of which are 0. We call

$$\sum n_i = \sum_P n_P$$

the **degree** of $\mathfrak{a}$, and we call n_i the **order** of $\mathfrak{a}$ at P_i.

If $x \in K$ and $x \neq 0$, then there is only a finite number of points P such that $\text{ord}_P x \neq 0$. Indeed, if x is constant, then $\text{ord}_P(x) = 0$ for all P. If x is not constant, then there is one point of $k(x)$ at which x has a zero, and one point at which x has a pole. Each of these points extends to only a finite number of points of K, which is a finite extension of $k(x)$. Hence we can associate a divisor with x, namely

$$(x) = \sum n_P P$$

where $n_P = \text{ord}_P(x)$. Divisors $\mathfrak{a}$ and $\mathfrak{b}$ are said to be **linearly equivalent** if $\mathfrak{a} - \mathfrak{b}$ is the divisor of a function. If $\mathfrak{a} = \sum n_P P$ and $\mathfrak{b} = \sum m_P P$ are divisors, we write

$$\mathfrak{a} \geq \mathfrak{b} \quad \text{if and only if} \quad n_P \geq m_P \quad \text{for all} \quad P.$$

This clearly defines a (partial) ordering among divisors. We call $\mathfrak{a}$ **positive** if $\mathfrak{a} \geq 0$.

If $\mathfrak{a}$ is a divisor, we denote by $L(\mathfrak{a})$ the set of all elements $x \in K$ such that $(x) \geq -\mathfrak{a}$. If $\mathfrak{a}$ is a positive divisor, then $L(\mathfrak{a})$ consists of all the functions in K which have poles only in $\mathfrak{a}$, with multiplicities at most those of $\mathfrak{a}$. It is clear that $L(\mathfrak{a})$ is a vector space over the constant field k for any divisor $\mathfrak{a}$. We let $l(\mathfrak{a})$ be its dimension.

Our main purpose is to investigate more deeply the dimension $l(\mathfrak{a})$ of the vector space $L(\mathfrak{a})$ associated with a divisor $\mathfrak{a}$ of the curve (we could say of the function field).

Let P be a point of V, and $\mathfrak{o}$ its local ring in K. Let $\mathfrak{p}$ be its maximal ideal. Since k is algebraically closed, $\mathfrak{o}/\mathfrak{p}$ is canonically isomorphic to k. We know that $\mathfrak{o}$ is a valuation ring, belonging to a discrete valuation. Let t be a generator of the maximal ideal. Let x be an element of $\mathfrak{o}$. Then for some constant a_0 in k, we can write $x \equiv a_0 \bmod \mathfrak{p}$. The function $x - a_0$ is in $\mathfrak{p}$, and has a zero at $\mathfrak{o}$. We can therefore write $x - a_0 = t y_0$, where y_0 is in $\mathfrak{o}$. Again by a similar argument we get $y_0 = a_1 + t y_1$ with $y_1 \in \mathfrak{o}$, and

$$x = a_0 + a_1 t + y_1 t^2.$$

Continuing this procedure, we obtain an expansion of x into a power series,

$$x = a_0 + a_1 t + a_2 t^2 + \cdots.$$

It is trivial that if each coefficient a_i is equal to 0, then $x = 0$.

The quotient field K of $\mathfrak{o}$ can be embedded in the power series field $k((t))$

as follows. If x is in K, then for some power t^s, the function $t^s x$ lies in $\mathfrak{o}$, and hence x can be written

$$x = \frac{a_{-s}}{t^s} + \cdots + \frac{a_{-1}}{t} + a_0 + a_1 t + \cdots.$$

If u is another generator of $\mathfrak{p}$, then clearly $k((t)) = k((u))$, and our power series field depends only on P. We denote it by K_P. An element ξ_P of K_P can be written $\xi_P = \sum_{v=m}^{\infty} a_v t^v$ with $a_m \neq 0$. If $m < 0$, we say that ξ_P has a **pole** of order $-m$. If $m > 0$ we say that ξ_P has a **zero** of order m, and we let $m = \text{ord}_P \xi_P$.

Lemma. *For any divisor $\mathfrak{a}$ and any point P, we have*

$$l(\mathfrak{a} + P) \leq l(\mathfrak{a}) + 1,$$

and $l(\mathfrak{a})$ is finite.

Proof. If $\mathfrak{a} = 0$ then $l(\mathfrak{a}) = 1$ and $L(\mathfrak{a})$ is the constant field because a function without poles is constant. Hence if we prove the stated inequality, it follows that $l(\mathfrak{a})$ is finite for all $\mathfrak{a}$. Let m be the multiplicity of P in $\mathfrak{a}$. Suppose there exists a function $z \in L(\mathfrak{a} + P)$ but $z \notin L(\mathfrak{a})$. Then

$$\text{ord}_P x = -(m + 1).$$

Let $w \in L(\mathfrak{a} + P)$. Looking at the leading term of the power series expansion at P for w, we see that there exists a constant c such that $w - cz$ has order $\geq -m$ at P, and hence $w \in L(\mathfrak{a})$. This proves the inequality, and also the lemma.

Let A^* be the cartesian product of all K_P, taken over all points P. An element of A^* can be viewed as an infinite vector $\xi = (\ldots, \xi_P, \ldots)$ where ξ_P is an element of K_P. The selection of such an element in A^* means that a random power series has been selected at each point P. Under componentwise addition and multiplication, A^* is a ring. It is too big for our purposes, and we shall work with the subring A consisting of all vectors such that ξ_P has no pole at P for all but a finite number of P. This ring A will be called the ring of **adeles**. Note that our function field K is embedded in A under the mapping

$$x \mapsto (\ldots, x, x, x, \ldots),$$

i.e., at the P-component we take x viewed as a power series in K_P. In particular, the constant field k is also embedded in A, which can be viewed as an algebra over k (infinite dimensional).

Let $\mathfrak{a}$ be a divisor on our curve. We shall denote by $\Lambda(\mathfrak{a})$ the subset of A consisting of all adeles ξ such that $\text{ord}_P \, \xi_P \geq -\text{ord}_P \, \mathfrak{a}$. Then $\Lambda(\mathfrak{a})$ is immediately seen to be a k-subspace of A. The set of all such $\Lambda(\mathfrak{a})$ can be taken as a fundamental system of neighborhoods of 0 in A, and define a topology in A which thereby becomes a topological ring.

The set of functions x such that $(x) \geq -\mathfrak{a}$ is our old vector space $L(\mathfrak{a})$, and is immediately seen to be equal to $\Lambda(\mathfrak{a}) \cap K$.

Let $\mathfrak{a}$ be a divisor, $\mathfrak{a} = \Sigma \, n_i P_i$, and let $\Sigma \, n_i$ be its degree. The purpose of this chapter is to show that $\deg(\mathfrak{a})$ and $l(\mathfrak{a})$ have the same order of magnitude, and to get precise information on $l(\mathfrak{a}) - \deg(\mathfrak{a})$. We shall eventually prove that there is a constant g depending on our field K alone such that

$$l(\mathfrak{a}) = \deg(\mathfrak{a}) + 1 - g + \delta(\mathfrak{a}),$$

where $\delta(\mathfrak{a})$ is a non-negative integer, which is 0 if $\deg(\mathfrak{a})$ is sufficiently large ($> 2g - 2$).

We now state a few trivial formulas on which we base further computations later. If B and C are two k-subspaces of A, and $B \supset C$, then we denote by $(B : C)$ the dimension of the factor space B mod C over k.

Proposition 2.1. *Let $\mathfrak{a}$ and $\mathfrak{b}$ be two divisors. Then $\Lambda(\mathfrak{a}) \supset \Lambda(\mathfrak{b})$ if and only if $\mathfrak{a} \geq \mathfrak{b}$. If this is the case, then*

1. $(\Lambda(\mathfrak{a}) : \Lambda(\mathfrak{b})) = \deg(\mathfrak{a}) - \deg(\mathfrak{b})$, *and*
2. $(\Lambda(\mathfrak{a}) : \Lambda(\mathfrak{b})) = ((\Lambda(\mathfrak{a}) + K) : (\Lambda(\mathfrak{b}) + K))$
$$+ ((\Lambda(\mathfrak{a}) \cap K) : (\Lambda(\mathfrak{b}) \cap K)).$$

Proof. The first assertion is trivial. Formula 1 is easy to prove as follows. If a point P appears in $\mathfrak{a}$ with multiplicity d and in $\mathfrak{b}$ with multiplicity e, then $d \geq e$. If t is an element of order 1 at P in K_P, then the index $(t^{-d}K_P : t^{-e}K_P)$ is obviously equal to $d - e$. The index in formula 1 is clearly the sum of the finite number of local indices of the above type, as P ranges over all points in $\mathfrak{a}$ or $\mathfrak{b}$. This proves formula 1. As to formula 2, it is an immediate consequence of the elementary homomorphism theorems for vector spaces, and its formal proof will be left as an exercise to the reader.

From Proposition 2.1 we get a fundamental formula:

(1) $\deg(\mathfrak{a}) - \deg(\mathfrak{b}) = (\Lambda(\mathfrak{a}) + K : \Lambda(\mathfrak{b}) + K) + l(\mathfrak{a}) - l(\mathfrak{b})$

for two divisors $\mathfrak{a}$ and $\mathfrak{b}$ such that $\mathfrak{a} \geq \mathfrak{b}$. For the moment we cannot yet separate the middle index into two functions of $\mathfrak{a}$ and $\mathfrak{b}$, because we do not know that $(A : \Lambda(\mathfrak{b}) + K)$ is finite. This will be proved later.

Let y be a non-constant function in K. Let $\mathfrak{c}$ be the divisor of its poles, and write $\mathfrak{c} = \Sigma \, e_i P_i$. The points P_i in $\mathfrak{c}$ all induce the same point Q of the rational curve having function field $k(y)$, and the e_i are by definition the ramification indices of the discrete value group in $k(y)$ associated with the point Q, and the extensions of this value group to K. These extensions correspond to the points P_i. We shall now prove that the degree $\Sigma\ e_i$ of c is equal to $[K : k(y)]$. We denote $[K : k(y)]$ by n.

Let $z_1, \ldots, z_n$ be a linear basis of K over $k(y)$. After multiplying each z_j with a suitable polynomial in $k[y]$ we may assume that they are integral over $k[y]$, i.e., that no place of K which is finite on $k[y]$ is a pole of any z_j. All the poles of the z_j are therefore among the P_i above appearing in c. Hence there is an integer μ_0 such that $z_j \in L(\mu_0 c)$. Let μ be a large positive integer. For any integer s satisfying $0 \le s \le \mu - \mu_0$ we get therefore

$$y^s z_j \in L(\mu c),$$

and so $l(\mu c) \ge (\mu - \mu_0 + 1)n$.

Let N_μ be the integer $(\Lambda(\mu c) + K : \Lambda(0) + K)$, so $N_\mu \ge 0$. Putting $b = 0$ and $a = \mu c$ in the fundamental formula (1), we get

$$\mu\left(\sum e_i\right) = N_\mu + l(\mu c) - 1$$

(2)
$$\ge N_\mu + (\mu - \mu_0 + 1)n - 1.$$

Dividing (2) by μ and letting μ tend to infinity, we get $\Sigma\ e_i \ge n$. Taking into account the corollary to Theorem 1.4 we get

Theorem 2.2. *Let K be the function field of a curve, and $y \in K$ a nonconstant function. If c is the divisor of poles of y, then $\deg(c) = [K : k(y)]$. Hence the degree of a divisor of a function is equal to 0 (a function has as many zeros as poles).*

Proof. If we let c' be the divisor of zeros of y then c' is the divisor of poles of $1/y$, and $[K : k(1/y)] = n$ also.

Corollary. $\deg(a)$ *is a function of the linear equivalence class of a.*

A function depending only on linear equivalence will be called a class function. We see that the degree is a class function.

Returning to (2), we can now write

$$\mu n \ge N_\mu + \mu n - \mu_0 n + n - 1$$

whence

$$N_\mu \le \mu_0 n - n + 1$$

and this proves that N_μ is uniformly bounded. Hence for large μ,

$$N_\mu = (\Lambda(\mu c) + K : \Lambda(0) + K)$$

is constant, because it is always a positive integer.

Now define a new function of divisors, $r(a) = \deg(a) - l(a)$. Both $\deg(a)$ and $l(a)$ are class functions, the former by Theorem 2.2 and the latter because the map $z \mapsto yz$ for $z \in L(a)$ is a k-isomorphism between $L(a)$ and $L(a - (y))$.

The fundamental formula (1) can be rewritten

(3) $$0 \leq (\Lambda(a) + K : \Lambda(b) + K) = r(a) - r(b)$$

for two divisors a and b such that $a \geq b$. Put $b = 0$ and $a = \mu c$, so

$$(\Lambda(\mu c) + K : \Lambda(0) + K) = r(\mu c) - r(0).$$

This and the result of the preceding paragraph show that $r(\mu c)$ is uniformly bounded for all large μ.

Let b now be any divisor. Take a function $z \in k[y]$ having high zeros at all points of b except at those in common with c (i.e., poles of y). Then for some μ, $(z) + \mu c \geq b$. Putting $a = \mu c$ in (3) above, and using the fact that $r(a)$ is a class function, we get

$$r(b) \leq r(\mu c)$$

and this proves that for an arbitrary divisor b the integer $r(b)$ is bounded. (The whole thing is of course pure magic.) This already shows that $\deg(b)$ and $l(b)$ have the same order of magnitude. We return to this question later. For the moment, note that if we now keep b fixed, and let a vary in (3), then $\Lambda(a)$ can be increased so as to include any element of A. On the other hand, the index in that formula is bounded because we have just seen that $r(a)$ is bounded. Hence for some divisor a it reaches its maximum, and for this divisor a we must have $A = \Lambda(a) + K$. We state this as a theorem.

Theorem 2.3. *There exists a divisor a such that $A = \Lambda(a) + K$. This means that the elements of K can be viewed as a lattice in A, and that there is a neighborhood $\Lambda(a)$ which when translated along all points of this lattice covers A.*

This result allows us to split the index in (1). We denote the dimension of $(A : \Lambda(a) + K)$ by $\delta(a)$. We have just proved that it is finite, and (1) becomes

(4) $$\deg(a) - \deg(b) = \delta(b) - \delta(a) + l(a) - l(b)$$

or in other words

(5) $\qquad l(\mathfrak{a}) - \deg(\mathfrak{a}) - \delta(\mathfrak{a}) = l(\mathfrak{b}) - \deg(\mathfrak{b}) - \delta(\mathfrak{b}).$

This holds for $\mathfrak{a} \geqq \mathfrak{b}$. However, since two divisors have a sup, (5) holds for any two divisors $\mathfrak{a}$ and $\mathfrak{b}$. The **genus** of K is defined to be that integer g such that

$$l(\mathfrak{a}) - \deg(\mathfrak{a}) - \delta(\mathfrak{a}) = 1 - g.$$

It is an invariant of K. Putting $\mathfrak{a} = 0$ in this definition, we see that $g = \delta(0)$, and hence that g is an integer $\geqq 0$, $g = (A : \Lambda(0) + K)$. Summarizing, we have

Theorem 2.4. *There exists an integer $g \geqq 0$ depending only on K such that for any divisor $\mathfrak{a}$ we have*

$$l(\mathfrak{a}) = \deg(\mathfrak{a}) + 1 - g + \delta(\mathfrak{a}),$$

where $\delta(\mathfrak{a}) \geqq 0$.

By a **differential** λ of K we shall mean a k-linear functional of A which vanishes on some $\Lambda(\mathfrak{a})$, and also vanishes on K (considered to be embedded in A). The first condition means that λ is required to be continuous, when we take the discrete topology on k. Having proved that $(A : \Lambda(\mathfrak{a}) + K)$ is finite, we see that a differential vanishing on $\Lambda(\mathfrak{a})$ can be viewed as a functional on the factor space

$$A \mod \Lambda(\mathfrak{a}) + K;$$

and that the set of such differentials is the dual space of our factor space, its dimension over k being therefore $\delta(\mathfrak{a})$.

Note in addition that the differentials form a vector space over K. Indeed, if λ is a differential vanishing on $\Lambda(\mathfrak{a})$, if ξ is an element of A, and y an element of K, we can define $y\lambda$ by $(y\lambda)(\xi) = \lambda(y\xi)$. The functional $y\lambda$ is again a differential, for it clearly vanishes on K, and in addition, it vanishes on $\Lambda(\mathfrak{a} + (y))$.

We shall call the sets $\Lambda(\mathfrak{a})$ **parallelotopes**. We then have the following theorem.

Theorem 2.5. *If λ is a differential, there is a maximal parallelotope $\Lambda(\mathfrak{a})$ on which λ vanishes.*

Proof. If λ vanishes on $\Lambda(\mathfrak{a}_1)$ and $\Lambda(\mathfrak{a}_2)$, and if we put

$$\mathfrak{a} = \sup (\mathfrak{a}_1, \mathfrak{a}_2)$$

then λ vanishes on $\Lambda(\alpha)$. Hence to prove our theorem it will suffice to prove that the degree of α is bounded. If $y \in L(\alpha)$, so $(y) \geq -\alpha$, then $y\lambda$ vanishes on $\Lambda(\alpha + y))$ which contains $\Lambda(0)$ because $\alpha + (y) \geq 0$. If $y_1, \ldots, y_n$ are linearly independent over k, then so are $y_1 \lambda, \ldots, y_n \lambda$. Hence we get

$$\delta(0) \geq l(\alpha) = \deg\alpha + 1 - g + \delta(\alpha).$$

Since $\delta(\alpha) \geq 0$, it follows that

$$\deg\alpha \leq \delta(0) + g - 1,$$

which proves the desired bound.

Theorem 2.6. *The differentials form a 1-dimensional K-space.*

Proof. Suppose we have two differentials λ and μ which are linearly independent over K. Suppose $x_1, \ldots, x_n$ and $y_1, \ldots, y_n$ are two sets of elements of K which are linearly independent over k. Then the differentials $x_1 \lambda, \ldots, x_n \lambda, y_1 \mu, \ldots, y_n \mu$ are linearly independent over k, for otherwise we would have a relation

$$\sum a_i x_i \lambda + \sum b_i y_i \mu = 0.$$

Letting $x = \Sigma \, a_i x_i$ and $y = \Sigma \, b_i y_i$, we get $x\lambda + y\mu = 0$, contradicting the independence of λ, μ over K.

Both λ and μ vanish on some parallelotope $\Lambda(\alpha)$, for if λ vanishes on $\Lambda(\alpha_1)$ and μ vanishes on $\Lambda(\alpha_2)$, we put $\alpha = \inf(\alpha_1, \alpha_2)$, and

$$\Lambda(\alpha) = \Lambda(\alpha_1) \cap \Lambda(\alpha_2).$$

Let $\mathfrak{b}$ be an arbitrary divisor. If $y \in L(\mathfrak{b})$, so that $(y) \geq -\mathfrak{b}$, then $y\lambda$ vanishes on $\Lambda(\alpha + (y))$ which contains $\Lambda(\alpha - \mathfrak{b})$ because $\alpha + (y) \geq \alpha - \mathfrak{b}$. Similarly, $y\mu$ vanishes on $\Lambda(\alpha - \mathfrak{b})$ and by definition and the remark at the beginning of our proof, we conclude that

$$\delta(\alpha - \mathfrak{b}) \geq 2l(\mathfrak{b}).$$

Using Theorem 2.4, we get

$$l(\alpha - \mathfrak{b}) - \deg(\alpha) + \deg(\mathfrak{b}) - 1 + g \geq 2l(\mathfrak{b})$$

$$\geq 2 \, (\deg(\mathfrak{b}) + 1 - g + \delta(\mathfrak{b}))$$

$$\geq 2 \deg(\mathfrak{b}) + 2 - 2g.$$

If we take $\mathfrak{b}$ to be a positive divisor of very large degree, then $L(\mathfrak{a} - \mathfrak{b})$ consists of 0 alone, because a function cannot have more zeros than poles. Since $\deg(\mathfrak{a})$ is constant in the above inequality, we get a contradiction, and thereby prove the theorem.

If λ is a non-zero differential, then all differentials are of type $y\lambda$. If $\Lambda(\mathfrak{a})$ is the maximal parallelotope on which λ vanishes, then clearly $\Lambda(\mathfrak{a} + (y))$ is the maximal parallelotope on which $y\lambda$ vanishes. We get therefore a linear equivalence class of divisors: if we define the **divisor** (λ) associated with λ to be $\mathfrak{a}$, then the divisor associated with $y\lambda$ is $\mathfrak{a} + (y)$. This divisor class is called the **canonical class** of K, and a divisor in it is called a **canonical divisor**.

Theorem 2.6 allows us to complete Theorem 2.4 by giving more information on $\delta(\mathfrak{a})$: we can now state the complete Riemann-Roch theorem.

Theorem 2.7. *Let $\mathfrak{a}$ be an arbitrary divisor of K. Then*

$$l(\mathfrak{a}) = \deg(\mathfrak{a}) + 1 - g + l(\mathfrak{c} - \mathfrak{a}).$$

where $\mathfrak{c}$ is any divisor of the canonical class. In other words,

$$\delta(\mathfrak{a}) = l(\mathfrak{c} - \mathfrak{a}).$$

Proof. Let $\mathfrak{c}$ be the divisor which is such that $\Lambda(\mathfrak{c})$ is the maximal parallelotope on which a non-zero differential λ vanishes. If $\mathfrak{b}$ is an arbitrary divisor and $y \in L(\mathfrak{b})$, then we know that $y\lambda$ vanishes on $\Lambda(\mathfrak{c} - \mathfrak{b})$. Conversely, by Theorem 2.6, any differential vanishing on $\Lambda(\mathfrak{c} - \mathfrak{b})$ is of type $z\lambda$ for some $z \in K$, and the maximal parallelotope on which $z\lambda$ vanishes is $(z) + \mathfrak{c}$, which must therefore contain $\Lambda(\mathfrak{c} - \mathfrak{b})$. This implies that

$$(z) \geqq -\mathfrak{b}, \quad \text{i.e.,} \quad z \in L(\mathfrak{b}).$$

We have therefore proved that $\delta(\mathfrak{c} - \mathfrak{b})$ is equal to $l(\mathfrak{b})$. The divisor $\mathfrak{b}$ was arbitrary, and hence we can replace it by $\mathfrak{c} - \mathfrak{a}$, thereby proving our theorem.

Corollary 1. *If $\mathfrak{c}$ is a canonical divisor, then $l(\mathfrak{c}) = g$.*

Proof. Put $\mathfrak{a} = 0$ in the Riemann-Roch theorem. Then $L(\mathfrak{a})$ consists of the constants alone, and so $l(\mathfrak{a}) = 1$. Since $\deg(0) = 0$, we get what we want.

Corollary 2. *The degree of the canonical class is $2g - 2$.*

Proof. Put $\mathfrak{a} = \mathfrak{c}$ in the Riemann-Roch theorem, and use Corollary 1.

Corollary 3. *If* $\deg(\alpha) > 2g - 2$, *then* $\delta(\alpha) = 0$.

Proof. $\delta(\alpha)$ is equal to $l(\mathfrak{c} - \alpha)$. Since a function cannot have more zeros than poles, $L(\mathfrak{c} - \alpha) = 0$ if $\deg(\alpha) > 2g - 2$.

§3. Remarks on Differential Forms

A **derivation** D of a ring R is a mapping $D: R \to R$ of R into itself which is linear and satisfies the ordinary rule for derivatives, i.e.,

$$D(x + y) = Dx + Dy, \quad \text{and} \quad D(xy) = xDy + yDx.$$

As an example of derivations, consider the polynomial ring $k[X]$ over a field k. For each variable X, the derivative $\partial/\partial X$ taken in the usual manner is a derivation of $k[X]$. We also get a derivation of the quotient field in the obvious manner, i.e., by defining $D(u/v) = (vDu - uDv)/v^2$.

We shall work with derivations of a field K. A derivation of K is **trivial** if $Dx = 0$ for all $x \in K$. It is trivial **over a subfield** k of K if $Dx = 0$ for all $x \in k$. A derivation is always trivial over the prime field: one sees that $D(1) = D(1 \cdot 1) = 2D(1)$, whence $D(1) = 0$.

We now consider the problem of extending a derivation D on K. Let $E = K(x)$ be generated by one element. If $f \in K[X]$, we denote by $\partial f/\partial x$ the polynomial $\partial f/\partial X$ evaluated at x. Given a derivation D on K, does there exist a derivation D^* on $K(x)$ coinciding with D on K? If $f(X) \in K[X]$ is a polynomial vanishing on x, then any such D^* must satisfy

$$(1) \qquad 0 = D^*(f(x)) = f^D(x) + \sum (\partial f/\partial x)D^*x,$$

where f^D denotes the polynomial obtained by applying D to all coefficients of f. Note that if relation (1) is satisfied for every element in a finite set of generators of the ideal in $K[X]$ vanishing on x, then (1) is satisfied by every polynomial of this ideal. This is an immediate consequence of the rules for derivations.

The above necessary condition for the existence of a D^* turns out to be sufficient.

Lemma 3.1. *Let D be a derivation of a field K. Let x be any element in an extension field of K, and let $f(X)$ be a generator for the ideal determined by x in $K[X]$. Then, if u is an element of $K(x)$ satisfying the equation*

$$0 = f^D(x) + f'(x)u,$$

there is one and only one derivation $D*$ of $K(x)$ coinciding with D on K, and such that $D*x = u$.

Proof. The necessity has been shown above. Conversely, if $g(x)$, $h(x)$ are in $K[x]$, and $h(x) \neq 0$, one verifies immediately that the mapping $D*$ defined by the formulas

$$D*g(x) = g^D(x) + g(x)u$$

$$D*(g/h) = \frac{hD*g - gD*h}{h^2}$$

is well defined and is a derivation of $K(x)$.

Consider the following special cases. Let D be a given derivation on K.

Case 1. x is separable algebraic over K. Let $f(X)$ be the irreducible polynomial satisfied by x over K. Then $f'(x) \neq 0$. We have

$$0 = f^D(x) + f'(x)u,$$

whence $u = -f^D(x)/f'(x)$. Hence D extends to $K(x)$ uniquely. If D is trivial on K, then D is trivial on $K(x)$.

Case 2. x is transcendental over K. Then D extends, and u can be selected arbitrarily in $K(x)$.

Case 3. x is purely inseparable over K, so $x^p - a = 0$, with $a \in K$. Then D extends to $K(x)$ if and only if $Da = 0$. In particular if D is trivial on K, then u can be selected arbitrarily.

From these three cases, we see that x is separable algebraic over K if and only if every derivation D of $K(x)$ which is trivial on K is trivial on $K(x)$. Indeed, if x is transcendental, we can always define a derivation trivial on K but not on x, and if x is not separable, but algebraic, then $K(x^p) \neq K(x)$, whence we can find a derivation trivial on $K(x^p)$ but not on $K(x)$.

The derivations of a field K form a vector space over K if we define zD for $z \in K$ by $(zD)(x) = zDx$.

Let K be a function field over the algebraically closed constant field k (function field means, as before, function field in one variable). It is an elementary matter to prove that there exists an element $x \in K$ such that K is separable algebraic over $k(x)$ (cf. *Algebra*). In particular, a derivation on K is then uniquely determined by its effect on $k(x)$.

We denote by $\mathfrak{D}$ the K-vector space of derivations D of K over k, (derivations of K which are trivial on k). For each $z \in K$, we have a pairing

$$(D, z) \mapsto Dz$$

of $(\mathfrak{D}, K)$ into K. Each element z of K gives therefore a K-linear functional of $\mathfrak{D}$. This functional is denoted by dz. We have

$$d(yz) = y\,dz + z\,dy$$
$$d(y + z) = dy + dz.$$

These linear functionals form a subspace $\mathcal{F}$ of the dual space of $\mathfrak{D}$, if we define $y\,dz$ by

$$(D, y\,dz) \mapsto yDz.$$

Lemma 3.2. *If K is a function field (in one variable) over the algebraically closed field k, then $\mathfrak{D}$ has dimension 1 over K. An element $t \in K$ is such that K over $k(t)$ is separable if and only if dt is a basis of the dual space of $\mathfrak{D}$ over K.*

Proof. If K is separable over $k(t)$, then any derivation on K is determined by its effect on t. If $Dt = u$, then $D = uD_1$, where D_1 is the derivation such that $D_1 t = 1$. Thus $\mathfrak{D}$ has dimension 1 over K, and dt is a basis of the dual space. On the other hand, using cases 2 and 3 of the extension theorem, we see at once that if K is not separable over $k(t)$, then $dt = 0$, and hence cannot be such a basis.

The dual space $\mathcal{F}$ of $\mathfrak{D}$ will be called the space of **differential forms** of K over k. Any differential form of K can therefore be written as $y\,dx$, where K is separable over $k(x)$.

§4. Residues in Power Series Fields

The results of this section will be used as lemmas to prove that the sum of the residues of a differential form in a function field of dimension 1 is 0.

Let $k((t))$ be a power series field, the field of coefficients being arbitrary (not necessarily algebraically closed). If u is an element of that field which can be written $u = a_1 t + a_2 t^2 + \cdots$ with $a_1 \neq 0$, then it is clear that

$$k((u)) = k((t)).$$

The order of an element of $k((t))$ can be computed in terms of u or of t. We call an element of order 1 a **local parameter** of $k((t))$.

Our power series field admits a derivation D_t defined in the obvious manner. Indeed, if $y = \Sigma\, a_\nu t^\nu$ is an element of $k((t))$ one verifies immediately that $D_t y = \Sigma\, \nu a_\nu t^{\nu-1}$ is a derivation. We sometimes denote $D_t y$ by dy/dt. There is also a derivation $D_u y$ defined in the same manner, and the classical chain rule $D_u y \cdot D_t u = D_t y$ (or better $dy/du \cdot du/dt = dy/dt$) holds here because it is a formal result.

If $y = \Sigma \, a_\nu t^\nu$, then a_{-1} (the coefficient of t^{-1}) is called the **residue of** y **with respect to** t, and denoted by $\mathrm{res}_t(y)$.

Proposition 4.1. *Let x and y be two elements of $k((t))$, and let u be another parameter of $k((t))$. Then*

$$\mathrm{res}_u\!\left(y \, \frac{dx}{du} \right) = \mathrm{res}_t\!\left(y \, \frac{dx}{dt} \right).$$

Proof. It clearly suffices to show that for any element y of $k((t))$ we have $\mathrm{res}_t(y) = \mathrm{res}_u(y \, dt/du)$. Since the residue is k-linear as a function of power series, and vanishes on power series which have a zero of high order, it suffices to prove our proposition for $y = t^n$ (n being an integer). Furthermore, our result is obviously true under the trivial change of parameter $t = au$, where a is a non-zero constant. Hence we may assume $t = u + a_2 u^2 + \cdots$, and $dt/du = 1 + 2a_2 u + \cdots$. We have to show that $\mathrm{res}_u(t^n \, dt/du) = 1$ when $n = -1$, and 0 otherwise.

When $n \geq 0$, the proposition is obvious, because $t^n \, dt/du$ contains no negative powers of t.

When $n = -1$, we have

$$\frac{1}{t} \frac{dt}{du} = \frac{1 + 2a_2 u + \cdots}{u + a_2 u^2 + \cdots} = \frac{1}{u} + \cdots,$$

and hence the residue is equal to 1, as desired.

When $n < -1$, we consider first the case in which the characteristic is 0. In this case, we have

$$\mathrm{res}_u(t^n \, dt/du) = \mathrm{res}_u\!\left(\frac{d}{du}\!\left(\frac{1}{n+1} t^{n+1} \right) \right)$$

and this is 0 for $n \neq -1$.

For arbitrary characteristic, and fixed $n < -1$, we have for $m > 1$,

$$\frac{1}{t^m} \frac{dt}{du} = \frac{1 + 2a_2 u + \cdots}{u^m(1 + a_3 u + \cdots)}$$

$$= \frac{F(a_2, a_3, \ldots)}{u} + \cdots,$$

where $F(a_2, a_3, \ldots)$ is a formal polynomial with integer coefficients. It is the same for all fields, and contains only a finite number of the coefficients

a_i. Hence the truth of our proposition is a formal consequence of the result in characteristic 0, because we have just seen that in that case, our polynomial $F(a_2, a_3, \ldots)$ is identically 0. This proves the proposition.

In view of Proposition 4.1, we shall call an expression of type $y dx$ (with x and y in the power series field) a **differential form of that field** and the **residue** $\text{res}(y dx)$ of that differential form is defined to be the residue $\text{res}_t(y dx/dt)$ taken with respect to any parameter t of our field.

We shall need a more general formula than that of Proposition 4. Given a power series field $k((u))$, let t be a non-zero element of that field of order $m \geq 1$. After multiplying t by a constant if necessary, we can write

$$t = u^m + b_1 u^{m+1} + b_2 u^{m+2} + \cdots$$
$$= u^m(1 + b_1 u + b_2 u^2 + \cdots).$$

Then the power series field $k((t))$ is contained in $k((u))$. In fact, one sees immediately that the degree of $k((u))$ over $k((t))$ is exactly equal to m. Indeed, by recursion, one can express any element y of $k((u))$ in the following manner

$$y = f_0(t) + f_1(t)u + \cdots + f_{m-1}(t)u^{m-1},$$

with $f_i(t) \in k((t))$. Furthermore, the elements $1, u, \ldots, u^{m-1}$ are linearly independent over $k((t))$, because our power series field $k((u))$ has a discrete valuation where u is an element of order 1, and t has order m. If we had a relation as above with $y = 0$, then two terms $f_i(t)u^i$ and $f_j(t)u^j$ would necessarily have the same absolute value with $i \neq j$. This obviously cannot be the case. Hence the degree of $k((u))$ over $k((t))$ is equal to m, and is equal to the ramification index of the valuation in $k((t))$ having t as an element of order 1 with respect to the valuation in $k((u))$ having u as element of order 1.

The following proposition gives the relations between the residues taken in $k((u))$ or in $k((t))$. By Tr we shall denote the trace from $k((u))$ to $k((t))$.

Proposition 4.2. *Let $k((u))$ be a power series field, and let t be a non-zero element of that field, of order $m \geq 1$. Let y be an element of $k((u))$. Then*

$$\text{res}_u\left(y \frac{dt}{du} du \right) = \text{res}_t (\text{Tr}(y) dt).$$

Proof. We have seen that the powers $1, u, \ldots, u^{m-1}$ form a basis for $k((u))$ over $k((t))$, and the trace of an element y of $k((u))$ can be computed from the matrix representing y on this basis. Multiplying t by a non-zero constant does not change the validity of the proposition. Hence we may assume that

$$t = u^m + b_1 u^{m+1} + \cdots = u^m(1 + b_1 u + b_2 u^2 + \cdots),$$

with $b_\nu \in k$. One can solve recursively

$$u^m = f_0(t) + f_1(t)u + \cdots + f_{m-1}(t)u^{m-1},$$

where $f_i(t)$ are elements of $k((t))$, and the coefficients of $f_i(t)$ are universal polynomials in $b_1, b_2, \ldots$, with **integer** coefficients, that is each $f_i(t)$ can be written

$$f_i(t) = \Sigma \, P_{i\nu}(b)t^\nu,$$

where each $P_{i\nu}(b)$ is a polynomial with integer coefficients, involving only a finite number of b's.

The matrix representing an arbitrary element

$$g_0(t) + g_1(t)u + \cdots + g_{m-1}(t)u^{m-1}$$

of $k((u))$ is therefore of type

$$\begin{pmatrix} G_{0,1}(t) & \cdots & G_{0,m-1}(t) \\ \cdots\cdots\cdots\cdots\cdots \\ G_{m-1,0}(t) & \cdots & G_{m-1,m-1}(t), \end{pmatrix}$$

where $G_{\nu\mu}(t) \in k((t))$, and where the coefficients of the $G_{\nu\mu}(t)$ are universal polynomials with integer coefficients in the b's and in the coefficients of the $g_j(t)$. This means that our formula, if it is true, is a formal identity having nothing to do with characteristic p, and that our verification can be carried out in characteristic 0.

This being the case, we can write $t = v^m$, where $v = u + c_2 u^2 + \cdots$ is another parameter of the field $k((u))$. This can be done by taking the binomial expansion for $(1 + b_1 u + \cdots)^{1/m}$. In view of Proposition 4.1, it will suffice to prove that

$$\mathrm{res}_v \left(y \, \frac{dt}{dv} \, dv \right) = \mathrm{res}_t(\mathrm{Tr}(y) \, dt).$$

By linearity, it suffices to prove this for $y = v^j$, $-\infty < j < +\infty$. (If y has a very high order, then both sides are obviously equal to 0, and y can be written as a sum involving a finite number of terms $a_j v^j$, and an element of very high order.)

If we write

$$j = ms + r \quad \text{with} \quad 0 \leq r \leq m - 1,$$

then $v^j = t^s v^r$ and $\text{Tr}(v^j) = t^s \text{Tr}(v^r)$. We have trivially

$$\text{Tr}(v^r) = \begin{cases} 0 & \text{if } r \neq 0 \\ m & \text{if } r \neq 0 \end{cases} \quad \text{whence} \quad \text{Tr}(v^j) = \begin{cases} mt^s & \text{if } j = ms \\ 0 & \text{otherwise.} \end{cases}$$

Consequently, we get

$$\text{res}_t(\text{Tr}(v^j)\, dt) = \begin{cases} m & \text{if } j = -m \\ 0 & \text{otherwise.} \end{cases}$$

On the other hand, our first expression in terms of v is equal to

$$\text{res}_v\, (v^j m v^{m-1}\, dv),$$

which is obviously equal to what we just obtained for the right-hand side. This proves our proposition.   .

In the preceding discussion, we started with a power series field $k((u))$ and a subfield $k((t))$. We conclude this section by showing that this situation is typical of power series field extensions.

Let $F = k((t))$ be a given power series field over an algebraically closed field k. We have a canonical k-valued place of F, mapping t on 0. Let E be a finite algebraic extension of F. Then the discrete valuation of F extends in at least one way to E, and so does our place. Let u be an element of E of order 1 at the extended valuation, which is discrete. If e is the ramification index, then we know by the corollary of Theorem 1 that $e \leq [E : F]$. We shall show that $e = [E : F]$ and hence that the extension of our place is unique.

An element y of E which is finite under the place has an expansion

$$y = a_0 + a_1 u + \cdots + a_{e-1} u^{e-1} + ty_1,$$

where y_1 is in E and is also finite. This comes from the fact that u^e and t have the same order in the extended valuation. Similarly, y_1 has also such an expansion, $y_1 = b_0 + b_1 u + \cdots + b_{e-1} u^{e-1} + ty_2$. Substituting this expression for y_1 above, and continuing the procedure, we see that we can write

$$y = f_0(t) + f_1(t)u + \cdots + f_{e-1}(t)u^{e-1},$$

where $f_i(t)$ is a power series in $k((t))$. Since the powers $1, u, \ldots, u^{e-1}$ are clearly linearly independent over $k((t))$, this proves that $e = [E : F]$, and that the extension of the place is unique.

Furthermore, to every element of E we can associate a power series in $\sum_{\mu=m}^{\infty} a_\mu u^\mu$, and one sees that every such power series arises from an element of E, because we can always replace u^e by ty, where y is finite under the place, and we can therefore solve recursively for a linear combination of the powers $1, u, \ldots, u^{e-1}$ with coefficients in $k((t))$. Summarizing, we get

Proposition 4.3. *Let $k((t))$ be a power series field over an algebraically closed field k. Then the natural k-valued place of $k((t))$ has a unique extension to any finite algebraic extension of $k((t))$. If E is such an extension, and u is an element of order 1 in the extended valuation, then E may be identified with the power series field $k((u))$ and $[E : F] = e$.*

§5. The Sum of the Residues

We return to global considerations, and consider a function field K of dimension 1 over an algebraically closed field k. The points P of K over k are identified with the k-valued places of K over k. For each such point, we have an embedding $K \to K_P$ of K into a power series field $k((t)) = K_P$ as in §2. Our first task will be to compare the derivations in K with the derivations in $k((t))$ discussed in §3.

Theorem 5.1. *Let y be an element of K. Let $t \in K$ be a local parameter at the point P, and let z be the element of K which is such that $dy = z\,dt$. If dy/dt is the derivative of y with respect to t taken formally from the power series expansion of y, then $z = dy/dt$.*

Proof. The statement of our theorem depends on the fact that every differential form of K can be written $z\,dt$ for some z, by §3. We know that K is separable algebraic over $k(t)$, and the irreducible polynomial equation $f(t, y) = 0$ of y over $k(t)$ is such that $f_y(t, y) \neq 0$. On the one hand, we have

$$0 = f_t(t, y)\,dt + f_y(t, y)\,dy,$$

whence $z = -f_t(t, y)/f_y(t, y)$ (cf. Lemma 2 of §3); and on the other hand, if we differentiate with respect to t the relation $f(t, y) = 0$ in the power series field, we get

$$0 = f_t(t, y) + f_y(t, y)\,\frac{dy}{dt}.$$

This proves our theorem.

Let ω be a differential form of K. Let P be a point of K, and t a local parameter, selected in K. Then we can write $\omega = y\,dt$ for some $y \in K$. Referring to Proposition 4.1 of §4, we can define the **residue of ω at P** to be the residue of $y\,dt$ at t, that is

$$\text{res}_P(\omega) = \text{res}_t(y).$$

If ω is written $x \, dz$ for $x, z \in K$, this residue is also written $\text{res}_P (x \, dz)$. We now state the main theorem of this section.

Theorem 5.2. *Let K be the function field of a curve over an algebraically closed constant field k. Let ω be a differential form of K. Then*

$$\sum_P \text{res}_P (\omega) = 0.$$

(The sum is taken over all points P, but is actually a finite sum since the differential form has only a finite number of poles.)

Proof. The proof is carried out in two steps, first in a rational function field, and then in an arbitrary function field using Proposition 4.2.

Consider first the case where $K = k(x)$, where x is a single transcendental quantity over k. The points P are in $1 - 1$ correspondence with the maps of x in k, and with the map $1/x \to 0$ (i.e., the place sending $x \to \infty$). If P is not the point sending x to infinity, but, say the point $x = a$, $a \in k$, then $x - a$ can be selected as parameter at P, and the residue of a differential form $y dx$ is the residue of y in its expansion in terms of $x - a$. The situation is the same as in complex variables.

We expand y into partial fractions,

$$y = \sum c_{\mu i}/(x - b_\mu)^i + f(x)$$

where $f(x)$ is a polynomial in $k[x]$. To get $\text{res}_P (y dx)$ we need consider only the coefficient of $(x - a)^{-1}$ and hence the sum of the residues taken over all P finite on x is equal to $\sum_\mu c_{\mu 1}$.

Now suppose P is the point at infinity. Then $t = 1/x$ is a local parameter, and $dx = -1/t^2 \, dt$. We must find the coefficient of $1/t$ in the expression $-y 1/t^2$. It is clear that the residue at t of $(-1/t^2) f(1/t)$ is equal to 0. The other expression can be expanded as follows:

$$-\frac{1}{t^2} \cdot \sum_{\mu, i} \frac{c_{\mu i}}{\left(\dfrac{1}{t} - b_\mu\right)^i} = -\sum_{\mu, i} c_{\mu i} t^{i-2} (1 + t b_\mu + \cdots)^i$$

and from this we get a contribution to the residue only from the first term, which gives precisely $-\sum c_{\mu 1}$. This proves our theorem in the case of a purely transcendental field.

Next, suppose we have a finite separable algebraic extension K of a purely transcendental field $F = k(x)$ of dimension 1 over the algebraically closed constant field k. Let Q be a point of F, and t a local parameter at Q in F. Let P be a point of K lying above Q, and let u be a local parameter at P in

K. Under the discrete valuation at P in K extending that of Q in F, we have $\text{ord}_P\, t = e \cdot \text{ord}_P\, u$. The power series field $k((u))$ is a finite extension of degree e of $k((t))$. Thus for each P we get an embedding of K in a finite algebraic extension of $k((t))$, and the place on K at P is induced by the canonical place of the power series field $k((u))$.

Let P_i $(i = 1, \ldots, s)$ be the points of K lying above Q. Let A be the algebraic closure of $k((t))$. The discrete valuation of $k((t))$ extends uniquely to a valuation of A, which is discrete on every subfield of A finite over $k((t))$ (Proposition 4.3 of §4). Suppose $K = F(y)$ is generated by one element y, satisfying the irreducible polynomial $g(Y)$ with leading coefficient 1 over F. It splits into irreducible factors over $k((t))$, say

(1) $$g(Y) = g_1(Y) \cdots g_r(Y)$$

of degrees d_j $(j = 1, \ldots, r)$. Let y_j be a root of $g_j(Y)$. Then the mapping $y \mapsto y_j$ induces an isomorphism of K into A. Two roots of the same g_j are conjugate over $k((t))$, and give rise to conjugate fields. By the uniqueness of the extension of the valuation ring, the induced valuation on K is therefore the same for two such conjugate embeddings. The ramification index relative to this embedding is d_j, and we see from (1) that $\Sigma\, d_j = n$. By Theorem 2.2 of §2 we now conclude that two distinct polynomials g_j give rise to two distinct valuations on K, and that $s = r$. We can therefore identify the fields $k((t))(y_i)$ with the fields K_{P_i}.

For each $i = 1, \ldots, r$ denote by Tr_i the trace from the field K_{P_i} to F_Q.

Proposition 5.3. *The notation being as above, let* Tr *be the trace from K to F. Then for any $y \in K$, we have*

$$\text{Tr}(y) = \sum_{i=1}^{r} \text{Tr}_i(y).$$

Proof. Suppose y is a generator of K over F. If $[K : F] = n$, then $\text{Tr}(y)$ is the coefficient of Y^{n-1} in the irreducible polynomial $g(Y)$ as above. A similar remark applies to the local traces, and our formula is then obvious from (1). If y is not a generator, let z be a generator. For some constant $c \in k$, $w = y + cz$ is a generator. The formula being true for cz and for w, and both sides of our equation being linear in y, it follows that the equation holds for y, as desired.

The next proposition reduces the theorem for an arbitrary function field K to a rational field $k(x)$.

Proposition 5.4. *Let k be algebraically closed. Let $F = k(x)$ be a purely transcendental extension of dimension 1, and K a finite algebraic separable extension of F. Let Q be a point of F, and P_i $(i = 1, \ldots, r)$ the* *points of K lying above Q. Let y be an element of K, and let* Tr *denote the trace from K to F. Then*

$$\operatorname{res}_Q (\operatorname{Tr}(y)\, dx) = \sum_{i=1}^{r} \operatorname{res}_{P_i} (y\, dx).$$

Proof. Let t be a local parameter at Q in $k(x)$, and let u_i be a local parameter in K at P_i. Let Tr_i denote the local trace from K_{P_i} to F_Q. We have

$$\sum \operatorname{res}_{P_i} (y\, dx) = \sum \operatorname{res}_{P_i} \left(y \frac{dx}{dt}\, dt \right)$$

$$= \sum \operatorname{res}_{P_i} \left(y \frac{dx}{dt} \frac{dt}{du_i}\, du_i \right)$$

and using Proposition 4.2 of §4, we see that this is equal to

$$\sum \operatorname{res}_Q \left(\operatorname{Tr}_i \left(y \frac{dx}{dt} \right) dt \right)$$

Since dx/dt is an element of $k(x)$, the trace is homogeneous with respect to this element, and the above expression is equal to

$$\sum \operatorname{res}_Q \left(\operatorname{Tr}_i(y) \frac{dx}{dt}\, dt \right) = \sum \operatorname{res}_Q (\operatorname{Tr}_i(y)\, dx)$$

$$= \operatorname{res}_Q \left(\sum \operatorname{Tr}_i(y)\, dx \right)$$

$$= \operatorname{res}_Q (\operatorname{Tr}(y)\, dx)$$

thereby proving our proposition.

Theorem 5.2 now follows immediately, because a differential form can be written $y\,dx$, where K is separable algebraic over $k(x)$.

Our theorem will allow us to identify differential forms of a function field K with the differentials introduced in §2, as k-linear functionals on the ring A of adeles which vanish on some $\Lambda(\mathfrak{a})$ and on K. This is done in the following manner. Let $\xi = (\ldots, \xi_P, \ldots)$ be an adele. Let $y\,dx$ be a differential form of K. Then the map

$$\lambda \colon \xi \mapsto \sum_P \operatorname{res}_P (\xi_P y\,dx)$$

is a k-linear map of A into k. Here, of course, in the expression $\operatorname{res}_P (\xi_P y\,dx)$, one views y and x as elements of K_P. It is also clear that all but a finite number of terms of our sum are 0.

Our k-linear map vanishes on some $\Lambda(\mathfrak{a})$, because the differential form has only a finite number of poles. Theorem 5.2 shows that it vanishes on K. It is therefore a differential, and in this way we obtain an embedding of the K-vector space of differential forms into the K-vector space of differentials. Since both spaces have dimension 1 over K (the latter by Theorem 2.6 of §2), this embedding is surjective.

Let ydx be a differential form of K. If P is a point of K, we can define the order of ydx at P easily. Indeed, let t be an element of order 1 at P. In the power series field $k((t))$, the element

$$y \frac{dx}{dt}$$

is a power series, with a certain order m_P independent of the chosen element t. We define m_P to be the order of ydx at P, and we let the **divisor of** ydx be

$$(ydx) = \sum m_P P.$$

Suppose $\operatorname{ord}_P(ydx) = m_P$. If $\operatorname{ord}_P(\xi_P) \geqq -m_P$, then

$$\operatorname{ord}_P(\xi_P ydx) \geqq 0,$$

and the residue $\operatorname{res}_P(\xi_P ydx)$ is 0. Hence the differential λ vanishes on $\Lambda(\mathfrak{a})$, where $\mathfrak{a} = (ydx)$. On the other hand, if $\Lambda(\mathfrak{b})$ is the maximal parallelotope on which λ vanishes, then $\Lambda(\mathfrak{b}) \supset \Lambda(\mathfrak{a})$, and $\mathfrak{b} \geqq \mathfrak{a}$. If $\mathfrak{b} > \mathfrak{a}$, then for some P, the coefficient of P in $\mathfrak{b}$ is $> m_P$, and hence the adele

$$(\ldots 0, 0, 1/t^{m_P+1}, 0, 0, \ldots)$$

lies in $\Lambda(\mathfrak{b})$. One sees immediately from the definitions that

$$\operatorname{res}_P(t^{-m_P-1} ydx) \neq 0,$$

and hence λ cannot vanish on $\Lambda(\mathfrak{b})$. Summarizing we have

Theorem 5.3. *Let K be a function field of dimension 1 over the algebraically closed constant field k. Each differential form ydx of K gives rise to a differential*

$$\lambda: \xi \mapsto \sum_P \operatorname{res}_P(\xi_P ydx),$$

and this induces a K-isomorphism of the K-space of differential forms onto

*the K-space of differentials. Furthermore, the divisors (ydx) and (λ) of §2
are equal.*

Corollary 1. *The integer* $\delta(\alpha)$ *is the dimension of the space of differential
forms* ω *such that*

$$(\omega) \geqq \alpha.$$

A differential form ω is said to be of **first kind** if it has no poles, that is if

$$(\omega) \geqq 0.$$

The space of differential forms of first kind is denoted by **dfk**. For any divisor
α, let $\text{Diff}(\alpha)$ be the space of differential forms ω such that $(\omega) \geqq -\alpha$. Then

$$\dim \text{Diff}(\alpha) = \delta(-\alpha).$$

If $\alpha \geqq 0$ it is clear that $\text{Diff}(\alpha)$ contains the space of differentials of first
kind.

Corollary 2. *For any divisisor* $\alpha > 0$ *we have*

$$\delta(-\alpha) = \deg\alpha - 1 + g$$

and

$$\dim \text{Diff}(\alpha)/\text{dfk} = \deg \alpha - 1.$$

Proof. Since $l(-\alpha) = 0$ because a function which has no poles and at least
one zero is identically 0, the formulas are special cases of the Riemann-Roch
theorem.

§6. The Genus Formula of Hurwitz

The formula compares the genus of a finite extension, in terms of the ramification indices.

Theorem 6.1. *Let k be algebraically closed, and let K be a function field
with k as constant field. Let E be a finite separable extension of K of degree
n. Let g_E and g_K be the genera of E and K respectively. For each point
P of K, and each point Q of E above P, assume that the ramification index
e_Q is prime to the characteristic of k. Then*

$$2g_E - 2 = n(2g_K - 2) + \sum_{Q} (e_Q - 1).$$

Proof. If ω is any non-zero differential form of K, then we know that its degree is $2g_K - 2$. Such a form can be written as ydx, with x, $y \in K$. We can also view x, y as elements of E; we can compute the degree in E, and compare it with that in K to get the formula, as follows. Let P be a point of K, and let t be a local parameter at P, that is an element of order 1 at P in K. If u is a local parameter at Q, then

$$t = u^e v,$$

where v is a unit at Q. Furthermore, $dt = u^e\, dv + eu^{e-1}v\, du$. Hence

$$\text{ord}_Q\,(ydx) = e_Q \cdot \text{ord}_P\,(ydx) + (e_Q - 1).$$

Summing over all Q over P, and then over all P yields the formula.

§7. Examples

Fields of genus 0. We leave to the reader as an exercise to prove that $k(x)$ itself has genus 0. Conversely, let K be a function field of genus 0 and let P be a point. By the Riemann-Roch theorem, there exists a non-constant function x in $L(P)$, because

$$l(P) = 1 + 1 - 0 + 0 = 2,$$

and the constants form a 1-dimensional subspace of $L(P)$. We contend that $K = k(x)$. Indeed, x has a pole of order 1 at P, and we know that $[K : k(x)]$ is equal to the degree of the divisor of poles, which is 1. Hence we see that K is the field of rational functions in x.

Fields of genus 1. Next let K be a function field of genus 1, and let P again be a point. We have $2g - 2 = 0$, so the Riemann-Roch theorem shows that the constant functions are the only elements of $L(P)$.

However, since $\deg(2P) = 2$, we have

$$l(2P) = 2 + 1 - 1 = 2,$$

so there exists a function x in K which has a pole of order 2 at P, and no other pole. Also

$$l(3P) = 3 + 1 - 1 = 3,$$

so there exists a function y in K which has a pole of order 3 at P. The seven functions 1, x, x^2, x^3, xy, y, y^2 must be linearly dependent because they all lie in $L(6P)$ and

$$l(6P) = 6 + 1 - 1 = 6.$$

In a relation of linear dependence, the coefficient of y^2 cannot be 0, for otherwise $y \in k(x)$ and this is impossible, as one sees from the parity of the poles of functions in $k(x)$ at P.

Since the degree of the divisor of poles of x is 2 we have

$$[K : k(x)] = 2.$$

Similarly,

$$[K : k(y)] = 3.$$

There is no strictly intermediate field between K and $k(x)$, and since y is not in $k(x)$, it follows that

$$K = k(x, y).$$

Furthermore, the relation of linear dependence between the above seven functions can be written

$$y^2 = c_1 y + c_2 xy + c_3 x^3 + c_4 x^2 + c_5 x + c_6.$$

In characteristic $\neq 2$ or 3, we can then make simple transformations of variables, and select x, y so that they satisfy the equation

$$y^2 = 4x^3 - c_2 x - c_3,$$

familiar from the theory of elliptic functions.

Hyperelliptic fields. Let $K = k(x, y)$ where y satisfies the equation

$$y^2 = f(x),$$

and $f(x)$ is a polynomial of degree n, which we may assume has distinct roots. Let us assume that the characteristic of k is $\neq 2$. Then the genus of K is

$$\left[\frac{n - 1}{2} \right].$$

Proof. Let $f(x) = \prod (x - a_i)$ where the elements a_i are distinct. Then K is unramified over $k(x)$ at all points except the points P_i corresponding to $x = a_i$, and also possibly at those points lying above $x = \infty$. At P_i the ramification index is 2. Suppose first that n is odd. Let $t = 1/x$ so that t has order 1 at ∞ in $k(x)$. We write

$$f(x) = t^{-n} \prod (1 - ta_i).$$

Each power series $1 - ta_i$ has a square root in $k[[t]]$, while for n odd, the square root of t^{-n} shows that $k(x, y)$ is ramified of order 2 at infinity. The Hurwitz genus formula yields

$$2g_K - 2 = 2(2 \cdot 0 - 2) + \sum_i (2 - 1) + (2 - 1) = -4 + n + 1.$$

Solving for g_K yields $g_K = (n - 1)/2$. If n is even, then the ramification index at infinity is 1 and the Hurwitz formula yields $g_K = (n - 2)/2$. This proves what we wanted.

§8. Differentials of Second Kind

In this section all fields are assumed of characteristic 0. A differential form ω is called of the **second kind** if it has no residues, that is if

$$\text{res}_P \, \omega = 0 \quad \text{for all} \quad P.$$

It is called of the **third kind** if its poles have order ≤ 1. The spaces dsk and dtk of such forms contain the differentials of first kind.

The Riemann-Roch theorem immediately shows that the differentials of first kind have dimension g, namely $\delta(0) = g$, equal to the genus.

A differential form is called **exact** if it is equal to dz for some function z. It is clear that an exact form is of the second kind. We shall be interested in the factor space

$$\text{dsk/exact}.$$

Theorem 8.1. *Assume that K has characteristic* 0. *Then*

$$\dim \text{dsk/exact} = 2g.$$

Proof. We define dsk(α) and dtk(α) just as we defined Diff(α), that is, forms of the prescribed kind whose divisor is $\geq -\alpha$.

Let $P_1, \ldots, P_r$ be distinct points, and let N be a positive integer such that $(N - 1)r > 2g - 2$. If a differential form is exact, say equal to dz, and lies in

$$\text{dsk}(N \sum P_i),$$

in other words, if it has poles at most at the points P_i of orders at most N, then $z \in L((N - 1) \Sigma P_i)$, and conversely. Note that

$$\text{dsk/exact} = \bigcup \text{dsk}(N \Sigma P_i)/dL ((N - 1) \Sigma P_i),$$

where the union is taken over all N as above, and all choices of points $P_1, \ldots, P_r$. To prove the theorem it will suffice to prove that each factor space on the right has dimension $2g$, because if that is the case, then increasing N or the set of points cannot yield any further contribution to dsk/exact.

This will then also prove:

Theorem 8.2. *Let $P_1, \ldots, P_r$ be distinct points, and let N be a positive integer such that $(N - 1)r > 2g - 2$. Then*

$$\text{dsk/exact} = \text{dsk}(N \Sigma P_i)/dL ((N - 1) \Sigma P_i).$$

Note that

$$\dim dL ((N - 1) \Sigma P_i) = l((N - 1) \Sigma P_i) - 1,$$

because the only functions z such that $dz = 0$ are the constants. On the other hand, also note that

$$\text{dfk} \cap \text{exact} = 0,$$

because a non-constant function z has a pole, and so dz also has a pole.

By Riemann-Roch (cf. Corollary 5.7) the dimension of the space of dtk having poles at most at the points P_i modulo the differentials of first kind has dimension

$$\delta(-\Sigma P_i) - g = r - 1.$$

Putting all this together, we find,

$$\dim \text{dsk} (N \Sigma P_i)/dL((N - 1) \Sigma P_i)$$

$$= \dim \text{Diff} (N \Sigma P_i)/dL((N - 1) \Sigma P_i) - \dim \text{dtk} (\Sigma P_i)$$

$$= \delta(-N \Sigma P_i) - [l((N - 1) \Sigma P_i) - 1] - (r - 1)$$

$$= 2g - 2 + Nr + 1 - g - (N - 1)r - 1 + g + 1 - (r - 1)$$
$$= 2g.$$

This proves the theorem.

§9. Function Fields and Curves

For technical simplicity, we assume again that k has characteristic zero.

Let K be a function field in one variable over a field k. This means that K is of transcendence degree 1, and finitely generated. If we can write $K = k(x, y)$, with two generators x, y, then we may call (x, y) the generic point of a plane curve, defined by the equation $f(X, Y) = 0$, if f is the irreducible polynomial vanishing on (x, y), determined up to a constant factor. A point (a, b) lies on the curve if and only if $f(a, b) = 0$. We shall say that the point is **simple** if $D_2 f(a, b) \neq 0$.

If $\mathfrak{o}$ is a discrete valuation ring of K over k (i.e., containing k) and $\mathfrak{m}$ its maximal ideal, then $\mathfrak{m}$ is principal, and any generator of $\mathfrak{m}$ is called a **local parameter** of $\mathfrak{o}$ or $\mathfrak{m}$. Assume that the residue class field $\mathfrak{o}/\mathfrak{m}$ is equal to k. Let $\varphi: \mathfrak{o} \to \mathfrak{o}/\mathfrak{m}$ be the canonical map. If $K = k(x, y)$ and $x, y \in \mathfrak{o}$, then we let $a = \varphi(x)$, $b = \varphi(y)$. We see that (a, b) is a point on the curve determined by (x, y). If $z \in K$, $z \notin \mathfrak{o}$, we can extend φ to all of K by letting $\varphi(z) = \infty$. We call φ a **place** of K (over k). We say that the point (a, b) is induced by the place on the curve.

Local Uniformization Theorem. *Let K be a function field in one variable over k.*

(1) *Let $K = k(x, y)$ where (x, y) satisfy an irreducible polynomial*

$$f(X, Y) = 0 \text{ over } k.$$

Let $a, b \in k$ be such that $f(a, b) = 0$ but $D_2 f(a, b) \neq 0$. Then there exists a unique place φ of K over k such that $\varphi(x) = a$, $\varphi(y) = b$, and if $\mathfrak{o}$ is the corresponding discrete valuation ring with maximal ideal $\mathfrak{m}$, then $x - a$ is a generator of $\mathfrak{m}$.

(2) *Conversely, let $\mathfrak{o}$ be a discrete valuation ring of K containing k, with maximal ideal, $\mathfrak{m}$, such that $\mathfrak{o}/\mathfrak{m} = k$. Let x be a generator of $\mathfrak{m}$. Then there exists $y \in \mathfrak{o}$ such that $K = k(x, y)$, and such that the point induced by the place on the curve is simple.*

Proof. To prove (1), we shall prove that any non-zero element

$$g(x, y) \in k[x, y]$$

can be written in the form

$$g(x, y) = (x - a)^{\nu} \frac{A(x, y)}{B(x, y)}$$

where A, B are polynomials, and $B(a, b) \neq 0$. This proves that the ring $\mathfrak{o}$ consisting of all quotients of polynomials $g_1(x, y)/g_2(x, y)$ with $g_2(a, b) \neq 0$ is a discrete valuation ring, and that $x - a$ is a generator of its maximal ideal.

If $g(a, b) \neq 0$, we are done, so assume $g(a, b) = 0$. Write

$$g(a, Y) = (Y - b)g_1(Y) \qquad g_1(Y) \in k[Y]$$
$$f(a, Y) = (Y - b)f_1(Y) \qquad f_1(Y) \in k[Y].$$

Then $f_1(b) \neq 0$ since $D_2 f(a, b) \neq 0$. Hence

$$g(a, Y)f_1(Y) = f(a, Y)g_1(Y).$$

It follows that

$$g(X, Y)f_1(Y) - f(X, Y)g_1(Y) = (X - a)A_1(X, Y)$$

for some polynomial A_1. Hence

$$g(x, y) = (x - a)A_1(x, y)/f_1(y).$$

If $A_1(a, b) \neq 0$, we are done. If not, we continue in the same way. We cannot continue indefinitely, for otherwise, we know that there exists some place of K over k inducing the given point, and $g(x, y)$ would have a zero of infinite order at the discrete valuation ring belonging to that place, which is impossible.

Conversely, to prove (2), let $K = k(x, z)$ where z is integral over $k[x]$. Let $z = z_1, \ldots, z_n$ $(n \geq 2)$ be the conjugates of z over $k(x)$, and extend $\mathfrak{o}$ to a valuation ring $\mathfrak{D}$ of $k(x, z_1, \ldots, z_n)$. Let

$$z = a_0 + a_1 x + \cdots + a_r x^r + \cdots$$

be the power series expansion of z, with $a_i \in k$, and let

$$P_r(x) = a_0 + \cdots + a_r x^r.$$

For $i = 1, \ldots, n$ let

$$y_i = \frac{z_i - P_r(x)}{x^r}.$$

If we take r large, then y_1 has no pole at $\mathfrak{D}$, but $y_2, \ldots, y_n$ have poles at $\mathfrak{D}$. The elements $y_1, \ldots, y_n$ are conjugate over $k(x)$. Let $f(X, Y)$ be the irreducible polynomial of (x, y) over k. Then

$$f(x, Y) = \psi_n(x) Y^n + \cdots + \psi_0(x).$$

Furthermore, $\psi_i(0) \neq 0$ for some i, otherwise we could factor out some power of X from $f(X, Y)$. We rewrite $f(x, Y)$ in the form

$$f(x, Y) = \psi_n(x) y_2 \ldots y_n (Y - y_1)\left(\frac{1}{y_2} Y - 1\right) \cdots \left(\frac{1}{y_n} Y - 1\right).$$

In the valuation determined by $\mathfrak{D}$, we see that the coefficient

$$u = \psi_n(x) y_2 \ldots y_n$$

cannot have a pole (otherwise divide the two expressions for $f(x, Y)$ by this coefficient and read the polynomial modulo the maximal ideal $\mathfrak{m}$ of $\mathfrak{o}$ to get a contradiction). If we denote by a bar the residue class of an element of $\mathfrak{o}$ mod $\mathfrak{m}$, then

$$0 \neq f(\bar{x}, Y) = (-1)^{n-1} \bar{u} (Y - \bar{y}_1).$$

By definition, $\bar{x} = a = 0$. We let $y = y_1$ and $\bar{y} = b$. Then

$$D_2 f(a, b) = (-1)^{n-1} \bar{u} \neq 0,$$

as was to be shown.

Corollary. *Let K be finite over $k(x)$. There is only a finite number of valuations of $k(x)$ over k which are ramified in K.*

Proof. There exist only a finite number of points (a, b) such that

$$f(a, b) = 0 \quad \text{and} \quad D_2 f(a, b) = 0,$$

and there exist only a finite number of valuation rings of K such that x does not lie in the valuation ring (i.e., such that x is at infinity).

If $K = k(x, y)$ and $f(x, y) = 0$ is the irreducible equation for x, y over k, then one calls the set of solutions (a, b) of the equation $f(a, b) = 0$ an affine plane curve, which is a model of the function field. If all its points are simple, the curve is called non-singular. The totality of all places of K (which are k-valued) is called the set of points on the complete non-singular curve associated with the function field K. There may of course not be a single model which is non-singular. Furthermore, an affine model always excludes "points at infinity", corresponding to places which map x to ∞. This could be taken care of by considering projective models of the function field. For our purposes, it suffices to call the set of all k-valued places the **complete non-singular model**, and to deal with this as the curve.

For our purposes, the genus of a curve is defined to be the genus of its function field.

Let R be the set of points of K. The elements of R are in bijection with the discrete valuation rings of K containing the constant field k. We speak of elements of R as the points on the above complete non-singular model. We write $R(K)$ if we wish to specify the reference to K.

Suppose that K is a finite extension of F. Then the inclusion $F \subset K$ gives rise to a mapping

$$\varphi: R(K) \to R(F)$$

which to each valuation ring $\mathfrak{o}$ of K associates the ring $\mathfrak{o} \cap F$ of F. This mapping φ can be representing on the points of an affine model. Indeed, if $K = k(x, y)$ as above, with irreducible equation $f(x, y) = 0$, and (a, b) is the point corresponding to $\mathfrak{o}$; and if $F = k(u, v)$, where u, v are in $\mathfrak{o}$ and the point induces the values (c, d) on (u, v), then we may write

$$u = \varphi_1(x, y), \qquad v = \varphi_2(x, y)$$

where φ_1, φ_2 are rational functions whose denominators do not vanish at (a, b). Then

$$c = \varphi_1(a, b) \qquad \text{and} \qquad d = \varphi_2(a, b).$$

§10. Divisor Classes

Let $\mathscr{D}_0$ be the group of divisors of degree 0, and $\mathscr{D}_l$ the subgroup of divisors of functions. The factor group

$$\mathscr{C} = \mathscr{D}_0/\mathscr{D}_l$$

is called the group of **divisor classes**.

Suppose that K is a finite extension of F, and let

$$\varphi: R(K) \to R(F)$$

be the associated map on the curves. Then φ induces a homomorphism

$$\varphi_*: \mathscr{C}(K) \to \mathscr{C}(F).$$

Indeed, φ_* is defined to be φ on points, and is extended by $\mathbf{Z}$-linearity to divisors. It is an elementary fact of algebra that φ_* maps divisors of functions to divisors of functions. In fact, if $z \in K$ then

$$\varphi_*((z)) = (N_{K/F}\, z),$$

in other words, the image of the divisor of (z) under φ is the divisor of the norm. For a proof, cf. Proposition 22 of Chapter I [La 2].

The map φ on divisors also induces a contravariant map

$$\varphi^*: \mathscr{C}(F) \to \mathscr{C}(K)$$

as follows. Given a point Q of F, let $P_1, \ldots, P_r$ be the distinct points of K lying above Q, and let e_i be the ramification index of P_i over Q. Then we define

$$\varphi^*(Q) = \sum_{i=1}^{r} e_i (P_i).$$

Then φ^* also maps the divisor of a function z in F to the divisor of that same function, viewed as element of K. This is obvious from the definition of the divisor of a function.

It is also immediate that

$$\varphi_* \varphi^* = [K : F],$$

because if $z \in F$ then $N_{K/F}(z) = z^n$ where $n = [K : F]$. Or, alternatively, because in the above notation,

$$\sum_{i=1}^{r} e_i = n.$$

CHAPTER II

The Fermat Curve

The purpose of this chapter is to give a significant example for the notions and theorems proved in the first chapter.

The reader interested in reaching the Abel-Jacobi as fast as possible can of course omit this chapter.

§1. The Genus

We consider the curve defined by the equation

$$x^N + y^N = 1$$

over an algebraically closed field k, and assume that N is prime to the characteristic of k. We denote this curve by $F(N)$ and call it the Fermat curve of level N. We suppose that $N \geq 3$ and again let K be its function field,

$$K = k(x, y).$$

We observe that the equation defining the curve is non-singular, so the discrete valuation rings in K are precisely the local rings of points on the curve, including the points with $x = \infty$, arising from the projective equation

$$x^N + y^N = z^N$$

with $z = 0$.

We have the expression

$$y^N = 1 - x^N = \prod (\zeta - x)$$

where the product is taken over all $\zeta \in \mu_N$ (N-th roots of unity). If x is set equal to an N-th root of unity, then we obtain a point on the Fermat curve with $y = 0$, and K is ramified of order N over this point, as is clear from the above equation. Hence

$$[K : k(x)] = N,$$

and the ramification index of K over the point $x = \zeta$ is N.

On the other hand, let $t = 1/x$. Then

$$y^N = \frac{1}{t^N} (t^N - 1),$$

and $-1 + t^N$ is a unit in $k[[t]]$. Hence $x = \infty$ (or $t = 0$) is not ramified in K, and there exist N distinct points of $F(N)$ (or K) lying over $x = \infty$, called the points at infinity *in this section*. If we put $z = ty$, then these N points have coordinates

$$z = \zeta \quad \text{for} \quad \zeta \in \mu_N.$$

By the Hurwitz genus formula, if we let g be the genus of $F(N)$, we get:

$$2g - 2 = -2N + \sum (e_p - 1) = -2N + N(N - 1).$$

Hence:

Theorem 1.1. *The genus of $F(N)$ is* $g = \dfrac{(N - 1)(N - 2)}{2}$.

§2. Differentials

For integers r, s such that $1 \leq r$, s we let

$$\omega_{r,s} = x^r y^s \frac{1}{N} \frac{dx^N}{x^N y^N}.$$

We can also write

$$\omega_{r,s} = x^{r-1} y^{s-1} \frac{dx}{y^{N-1}}.$$

Since

$$\frac{dx}{y^{N-1}} = -\frac{dy}{x^{N-1}},$$

we see that the only possible poles of $\omega_{r,s}$ lie among the points with $x = \infty$. In other words, if P is a point such that $x(P)$ is finite $\neq 0$ then the expression dy/x^{N-1} shows that $\omega_{r,s}$ has no pole at P. If P is a point such that $x(P) = 0$ then $y(P) \neq 0$, and the other expression shows that dx/y^{N-1} has no pole at P.

Theorem 2.1. *A basis for the differentials of first kind is given by $\omega_{r,s}$ with $1 \leq r,\ s$ such that $r + s \leq N - 1$.*

Proof. Suppose that $x(P) = \infty$. Put $t = 1/x$. Then

$$dx = -\frac{1}{t^2}\,dt,$$

and $\operatorname{ord}_P y = \operatorname{ord}_P x = 1$. Then

$$\omega_{r,s} = x^{r-1}y^{s-1}\left(\frac{-1}{t^2}\right)\frac{dt}{y^{N-1}},$$

from which it is clear that if $r + s \leq N - 1$ then $\operatorname{ord}_P \omega_{r,s} \geq 0$. This proves the theorem, because it is also clear that the differential forms as stated are linearly independent over the constants, and there are precisely g of them, where g is the genus of $F(N)$.

We observe that these forms have an additional structure. The group $\mu_N \times \mu_N$ acts as a group of automorphisms of $F(N)$ by the action

$$(x, y) \mapsto (\zeta_N^i x,\ \zeta_N^j y)$$

where ζ_N is a fixed primitive N-th root of unity. Over the complex numbers, we usually take $\zeta_N = e^{2\pi i/N}$. Then the form $\omega_{r,s}$ (without any restriction on the integers r, s) is an eigenform for the character $X_{r,s}$ such that

$$X_{r,s}(\zeta^i, \zeta^j) = \zeta^{ri+sj}.$$

The linear independence of the differentials of first kind in Theorem 2.1 can therefore also be seen from the fact that they are eigenforms for this Galois group, with distinct characters.

When we view $\mu_N \times \mu_N$ as a group of automorphisms of $F(N)$ we shall also write it as $G = G(N)$, and call it simply the group of natural automorphisms of the Fermat curve, or also the Galois group of $F(N)$ over $F(1)$.

Theorem 2.2. *The forms $\omega_{r,s}$ with*

$$1 \leq r,\ s \leq N - 1 \ and \ r + s \not\equiv 0 \bmod N$$

constitute a basis for dsk/exact.

Proof. Let

$$\infty = \sum_{i=1}^{N} (\infty_i)$$

be the divisor of points above $x = \infty$ on $F(N)$, taken with multiplicity 1
First we note that the space

$$dtk(\infty)/dfk$$

has dimension $N - 1$ by the Riemann-Roch theorem (Corollary 2 of Theorem 5.3, Chapter I). Checking the order of pole at infinity shows that the forms

$$\omega_{r,s} \quad \text{with} \quad r + s = N \quad \text{and} \quad 1 \leqq r, s$$

are of the third kind, and obviously linearly independent from the differentials of first kind in Theorem 2.1. Since they have the right dimension, they form a basis of $dtk(\infty)/dfk$.

Given any differential ω, it follows that there exists a homogeneous polynomial $h(x, y)$ of degree $N - 2$ such that

$$\omega - h(x, y) \frac{dx}{y^{N-1}}$$

is of the second kind. We apply this remark to the forms $\omega_{r,s}$ with $r + s \not\equiv 0$ mod N. We operate with (ζ, ζ) on the above difference, and note that the automorphisms of $F(N)$ preserve the spaces of dsk. Subtracting, we then find that

$$(1 - \zeta^{r+s})\omega_{r,s} \quad \text{with} \quad r + s \not\equiv 0 \bmod N$$

is of second kind, whence $\omega_{r,s}$ is of second kind.

Finally, we note that the forms $\omega_{r,s}$ with $1 \leqq r, s \leqq N - 1$ and $r + s \not\equiv 0$ mod N, taken modulo the exact forms are eigenforms for the Galois group $\mu_N \times \mu_N$ with distinct characters, and hence are linearly independent in dsk/exact. Since the number of such forms is precisely the dimension of dsk/exact, it follows that they form a basis for this factor space, thus proving the theorem.

§3. Rational Images of the Fermat Curve

Throughout this section, we let $1 \leqq r, s$ and $r + s \leqq N - 1$. Such a pair (r, s) will be called **admissible**. We shall consider rational images of the

Fermat curve (subfields of its function field) following Rohrlich [Ro] after
Faddeev [Fa 2].

We put

$$u = x^N \quad \text{and} \quad v = x^r y^s.$$

We let

$$D = \text{g.c.d.}(r, s, N) \quad \text{and} \quad M = N/D.$$

We write

$$r = r'D, \ s = s'D \quad \text{so that} \quad (r', s', M) = 1.$$

Then u, v are related by the equation

$$v^N = u^r (1 - u)^s,$$

which, in irreducible form, amounts to

$$v^M = u^{r'} (1 - u)^{s'}.$$

We let $F(r, s)$ be the "non-singular curve" whose function field is $k(u, v)$, so
that we have a map

$$F(N) \to F(r, s),$$

given in terms of coordinates by

$$(x, y) \mapsto (u, v) = (x^N, x^r y^s).$$

If $t \in \mathbf{Z}(M) = \mathbf{Z}/M\mathbf{Z}$, we let $\langle t \rangle_M$ be the integer such that

$$0 \leq \langle t \rangle_M \leq M - 1 \quad \text{and} \quad \langle t \rangle_M \equiv t \bmod M.$$

If $M = N$ we omit the subscript M from the notation. If we let $K(N)$ and
$K(r, s)$ be the function fields of $F(N)$ and $F(r, s)$ respectively, then $K(N)$ is
Galois over $K(r, s)$. Let $G(r, s)$ be the Galois group.

$$G \approx \mu_N \times \mu_N \begin{cases} \left. \begin{array}{c} K(N) = k(x, y) \\ | \\ K(r, s) = k(u, v) \end{array} \right\} G(r, s) \\ | \\ K(1) = k(u) \end{cases}$$

We note that $G(r, s)$ is the kernel of the character $X_{r,s}$, and that $K(r, s)$ over $K(1)$ is cyclic, of degree M. It is in fact a Kummer extension.

Special case. Consider the case when $N = p$ is prime ≥ 3 and

$$r = s = 1,$$

so the intermediate curve is defined by

$$v^p = u(1 - u),$$

which is therefore hyperelliptic. The change of variables

$$t = 2u - 1$$

changes this equation to

$$t^2 = 1 - 4v^p,$$

which is often easier to work with.

Let $m \in \mathbf{Z}(M)$. We say that m is (r, s)-**admissible** if $\langle mr \rangle$ and $\langle ms \rangle$ form an admissible pair, that is

$$1 \leq \langle mr \rangle, \langle ms \rangle \quad \text{and} \quad \langle mr \rangle + \langle ms \rangle \leq N - 1.$$

Theorem 3.1. *A basis of* dfk *on* $F(r, s)$ *is given by the forms*

$$\omega_{\langle mr \rangle, \langle ms \rangle}$$

for all (r, s)-*admissible elements* m.

Proof. It is clear that $x^{\langle mr \rangle} y^{\langle ms \rangle}$ lies in the function field of $F(r, s)$, and hence that the forms listed above are of the first kind on $F(r, s)$. Conversely, suppose ω is a dfk on $F(r, s)$. Write

$$\omega = \sum c_{q,t} \omega_{q,t}$$

where the sum is over all admissible pairs (q, t). Since $\omega_{q,t}$ is an eigenform of $\mu_N \times \mu_N$ with eigencharacter $X_{q,t}$, and since the factor group

$$(\mu_N \times \mu_N)/ \operatorname{Ker} X_{r,s}$$

is cyclic, it follows that if $c_{q,t} \neq 0$ then

$$X_{q,t} = X_{r,s}^m,$$

for some integer m. But then $q = \langle mr \rangle$ and $t = \langle ms \rangle$, as desired.

Suppose that $m \in \mathbf{Z}(M)^*$, and that m is (r, s)-admissible. Then the function fields of $F(r, s)$ and $F(\langle mr \rangle, \langle ms \rangle)$ are equal, for instance because

$$\text{Ker } X_{r,s} = \text{Ker } X_{\langle mr \rangle, \langle ms \rangle}.$$

The correspondence between the curves in terms of coordinates is given as follows.

Let $1 \leq m \leq M - 1$ be prime to M and such that its residue class mod M is (r, s)-admissible. Write

$$(\langle mr \rangle, \langle ms \rangle) = m(r, s) + N(i, j)$$

with some pair of integers i, j. Then we have a commutative diagram:

$$\begin{array}{ccc} & F(N) & \\ \swarrow & & \searrow \\ F(r, s) & \longrightarrow & F(\langle mr \rangle, \langle ms \rangle) \end{array}$$

where the bottom arrow is given by

$$(u, v) \mapsto (u, v^m u^i (1 - u)^j),$$

realizing the automorphism of the function field corresponding to the two models $F(r, s)$ and $F(\langle mr \rangle, \langle ms \rangle)$.

Two admissible pairs (r, s) and (q, t) are called **equivalent** if there exists $m \in \mathbf{Z}(M)^*$ such that $q = \langle mr \rangle$ and $t = \langle ms \rangle$. It is clear that inequivalent pairs correspond to distinct subfields $K(r, s)$ and $K(q, t)$. On the other hand, if (r, s) and (q, t) are equivalent, then

$$K(r, s) = K(q, t).$$

Given the admissible pair (r, s), and $m \in \mathbf{Z}(M)^*$, we observe that there is precisely one value of m or $-m$ such that $(\langle mr \rangle, \langle ms \rangle)$ is admissible. This follows at once from the fact that for any integer $a \not\equiv 0 \bmod N$ we have

$$\langle a \rangle + \langle -a \rangle = N.$$

We shall now apply this to the most interesting special case when $N = p$ is prime ≥ 3.

Theorem 3.2. *If $N = p$ is prime ≥ 3, then for every admissible pair (r, s) the curve $F(r, s)$ has genus $(p - 1)/2$, and $K(r, s) = K(1, s^*)$ for a uniquely determined integer s^* such that the pair $(1, s^*)$ is admissible.*

Proof. The genus can either be computed directly as we did for the Fermat curve, or one can use Theorem 3.1. The number of m such that $(\langle mr \rangle, \langle ms \rangle)$ is admissible is trivially computed to be $(p - 1)/2$, using the remark preceding the theorem. The statement that $K(r, s) = K(1, s^*)$ is clear.

Of course, instead of $(1, s^*)$ we could also have picked a representative in the equivalence class of (r, s) to be $(r^*, 1)$.

Remark. If we define $F(p - 1, 1)$ by

$$v^{p-1} = u^{p-1}(1 - u),$$

then $F(p - 1, 1)$ has genus 0, since it is also defined by

$$w^p = \frac{1 - u}{u},$$

where $w = v/u$, and $(1 - u)/u$ is a fractional linear transform of u, so a generator of $k(u)$. The function field of $F(p - 1, 1)$ is therefore equal to $k(w)$.

§4. Decomposition of the Divisor Classes

Let $F_k = F(1, k)$ for $k = 1, \ldots, p - 2$ and let

$$f_k : F(p) \to F_k$$

be the associated rational map. As we have seen in Chapter I, §10 there is an associated map f_{k*} on divisor classes, as well as an inverse map f_k^*. Let $\mathscr{C}_k = \mathscr{C}(F_k)$ be the group of divisor classes on F_k, and let

$$f = \oplus f_k : \mathscr{C} \to \bigoplus_{k=1}^{p-2} \mathscr{C}_k$$

be the direct sum of the f_k. Also let

$$f^* : \bigoplus_{k=1}^{p-2} \mathscr{C}_k \to \mathscr{C}$$

be the map

$$f* = \sum_{k=1}^{p-2} f_k^*.$$

We shall give the proof of the next theorem only over $k = \mathbf{C}$.

Theorem 4.1. *We have* $f* \circ f = p \cdot \mathrm{id}.$

Proof. Let A and B be the automorphisms of $F(N)$ induced by $(x, y) \mapsto$ $(\zeta x, y)$ and $(x, \zeta y)$ respectively. We use the same letters for the induced automorphisms of the divisor class group. For any divisor $\mathfrak{a}$ of degree 0 on $F(p)$ the elementary definitions of the direct image and inverse image give us the formula

$$f_k^* \circ f_{k_*}(\mathfrak{a}) = \sum_{j=0}^{p-1} (A^{-k}B)^j(\mathfrak{a}).$$

Hence

$$f* \circ f = \sum_{k=1}^{p-2} \sum_{j=0}^{p-1} A^{-kj} B^j.$$

We wish to show this is equal to $p \cdot \mathrm{id}$ on divisor classes. We need the fact that this is equivalent to showing this same relation when the map is viewed as being applied to differentials of first kind. Any general theory will prove this fact. The reader can deduce it for instance from the duality of Theorems 5.5 and 5.6 in Chapter IV. We assume this fact. Then it suffices to prove the desired relation when the map is applied to the differential forms $\omega_{r,s}$ with $r + s \leqq p - 1$. Since such forms are eigenforms for the Galois group of $F(p)$ over $F(1)$, we see that the above relation is equivalent to the relation

$$\sum_{k=1}^{p-2} \sum_{j=0}^{p-1} \zeta^{j(s-rk)} = p.$$

Since $r + s \leqq p - 1$ we see that for each pair r, s there exists a unique k satisfying

$$ks \equiv r \pmod{p}.$$

For this value of k, the sum over j is equal to p, and $p - p = 0$. For the other values of k, the sum over j is equal to 0. This proves the theorem.

The theorem gives us a sequence of maps

$$\mathscr{C} \xrightarrow{f} \bigoplus_{k=1}^{p-2} \mathscr{C}_k \xrightarrow{f*} \mathscr{C}$$

whose composition is $p \cdot \mathrm{id}$. In Chapter IV, we shall prove that the group of divisor classes is isomorphic to a complex torus, and it is clear that $f, f*$ are complex analytic homomorphisms, over the complex numbers. From Theorem 4.1 we conclude that $f^* \circ f$ has finite kernel, and from the fact that the dimension of the group of divisor classes is equal to the genus, we see that

$$\dim \mathscr{C} = \sum_{k=1}^{p-2} \dim \mathscr{C}_k.$$

Thus f must be surjective, and up to such a homomorphism with finite kernel, we have a decomposition of $\mathscr{C}$ into a direct sum of factors corresponding to the curves F_k with $k = 1, \ldots, p - 2$.

For more general results, cf. Koblitz-Rohrlich [Ko-R].

CHAPTER III

The Riemann Surface

The purpose of this chapter is to show how to give a structure of analytic manifold to the set of points on a curve in the complex numbers, but our treatment also applies to more general fields like p-adic numbers. We are principally interested in the complex case, in order to derive the Abel-Jacobi theorem in the next chapter.

§1. Topology and Analytic Structure

Assume now that k is locally compact. It can be shown that k is the real field, complex field, or a p-adic field. A point P of K is called k-**rational** if the residue class field $\mathfrak{o}/\mathfrak{p}$ of its valuation ring is equal to k itself. The local uniformization theorem shows in fact how to interpret primes as non-singular points on plane curves. We let R be the set of all k-rational points of K over k, and call it the **Riemann surface** of K over k. We can view the elements of K as functions on R. If $P \in R$, we denote by $\mathfrak{o}_P$ the valuation ring associated with P, and by $\mathfrak{m}_P$ its maximal ideal. If $z \in \mathfrak{o}_P$, then we define $z(P)$ to be the residue class of z mod $\mathfrak{m}_P$. Thus

$$z(P) \in k.$$

If $z \notin \mathfrak{o}_P$, then we define $z(P) = \infty$. The elements of k are constant functions, and are. in fact the only constant functions. If $z(P) = 0$, we say that z has a **zero** at P, and if $z(P) = \infty$, we say that z has a **pole** at P.

For each $x \in K$, we let $\Gamma_x = \{k, \infty\}$ be the Gauss sphere over k, that is, the compact space obtained by adjoining to k a point at infinity. We let

$$\Gamma = \prod_{x \in K} \Gamma_x.$$

We embed R in Γ in the obvious way: An element P goes on the product $\Pi\, x(P)$. We topologize R as a subspace of Γ, which amounts to saying that the topology is the one having the least amount of open sets such that all the functions $x \in K$ are continuous.

The product Γ is compact, and we contend that R is closed in Γ. This implies that R is compact.

Proof. Let $(a_x)_{x \in K}$ be in the closure. Let $\mathfrak{o}$ be the set of elements x in K such that $a_x \neq \infty$. Then $\mathfrak{o}$ is a valuation ring, whose corresponding point Q is such that $x(Q) = a_x$ for all x. This is easily proved. We first note that $k \subset \mathfrak{o}$ because $a(P) = a$ for all $a \in k$ and $P \in R$. Let $x, y \in \mathfrak{o}$. Then $a_x, a_y \neq \infty$. By assumption, there exists $P \in R$ such that $x(P)$, $y(P)$, and $(x + y)(P)$ are arbitrarily close to a_x, a_y, a_{x+y} respectively. For such P, we see that $x(P)$ and $y(P) \neq \infty$, whence $(x + y)(P) \neq \infty$. Hence $x + y \in \mathfrak{o}$. Similarly, $x - y$ and xy lie in $\mathfrak{o}$, which is therefore a ring. Furthermore, the map $x \mapsto a_x$ is a homormophism of $\mathfrak{o}$ into k, and is the identity on k. This follows from a continuity argument as above. Finally, $\mathfrak{o}$ is a valuation ring, for suppose $x \notin \mathfrak{o}$. Then $a_x = \infty$. Let $y = x^{-1}$. There exists $P \in R$ such that $x(P)$ is close to a_x and $y(P)$ is close to a_y. Since $x(P)$ is close to infinity, it follows that $y(P)$ is close to 0. Hence $a_y = 0$, so $y \in \mathfrak{o}$. This proves our assertion.

Let P be a point. Let t be a generator of the maximal ideal $\mathfrak{m}_P$, i.e. a local uniformizing parameter at P. We shall now prove that *the map*

$$Q \mapsto t(Q)$$

gives a topological isomorphism of a neighborhood of P onto a neighborhood of 0 in k.

According to the local uniformizing theorem, we can find generators t, y for K such that the point P is represented by a simple point with coordinates (a, b) in k, and in fact $a = 0$. Split the polynomial $f(0, Y)$ in the algebraic closure k^a of k. Then b is a root of multiplicity 1, so we have

$$f(0, Y) = (Y - b)(Y - b_2)^{e_2} \cdots (Y - b_r)^{e_r}.$$

The roots of a polynomial are continuous functions of the coefficients. There exists a neighborhood U of 0 in k such that for any element $\bar{t} \in U$, the polynomial $f(\bar{t}, Y)$ has exactly one root in k^a, with multiplicity 1, and this root is close to b (in the algebraic closure of k). However, using, for instance, the Newton approximation method, starting with the approximate root b, we can refine b to a root of $f(\bar{t}, Y)$ in k itself, if we took U sufficiently small. Hence the map $Q \mapsto t(Q)$ is injective on the set of Q such that $(t(Q), y(Q))$ lies in a suitably small neighborhood of $(0, b)$. Since the topology on R is determined by the functions in K, and since k is locally compact, we conclude that for U sufficiently small, the map $Q \mapsto t(Q)$ gives a topological isomorphism of U onto a neighborhood of 0 in k.

We can choose U such that $t(U)$ is a disc on the t-plane. We observe that if Q is close to P, then applying the local uniformization theorem shows that $t - t(Q)$ is a local uniformizing parameter at Q, because the condition concerning the second partial derivative is satisfied by continuity. Indeed, if $D_2 f(a, b) \neq 0$, then $D_2 f(a', b') \neq 0$ for all (a', b') sufficiently close to (a, b).

Theorem 1.1. *Taking as charts discs on t-planes as above, together with the maps given by local parameters, gives an analytic manifold structure to R.*

Proof. If $P \in R$, and t, u are two parameters at P, then t has a power series expansion in terms of u, say

$$t = \sum_{i=0}^{\infty} a_i u^i,$$

and since t is algebraic over $k(u)$, simple estimates show that this power series is convergent in some neighborhood of the origin. Hence

$$t(Q) = \sum a_i u(Q)^i$$

for Q close to 0 in the t-plane, and we see that the functions giving changes of charts are holomorphic.

Theorem 1 is valid for any field k which is real, complex, or p-adic. *From now on, we shall assume that $k = \mathbf{C}$ is the field of complex numbers.*

We have the notion of a meromorphic function on R, that is a quotient of holomorphic functions locally at each point. If f is such a function, we can write it locally around a point P as a power series

$$f(Q) = \sum b_i t(Q)^i$$

where t is a parameter at P, and a finite number of negative powers of t may occur. In this way, f may be viewed as embedded in the power series field $K_P \approx \mathbf{C}((t))$, and thus can be viewed as an adele because it has only a finite number of poles on R since R is compact.

Theorem 1.2. *Every meromorphic function on R is in K.*

Proof. Let L be the field of meromorphic functions on R. If $L \neq K$, then the degree $(L : K)_\mathbf{C}$ of the factor space of L mod K over the complex is infinite. We have:

$$(L : K)_\mathbf{C} = (L + \Lambda(0) : K + \Lambda(0))_\mathbf{C} + (L \cap \Lambda(0) : K \cap \Lambda(0))_\mathbf{C}.$$

The first term on the right is finite as was shown in Weil's proof of the Riemann-Roch theorem. The second is 0 because a function having no pole is a constant (by the maximum modulus principle). Contradiction.

Theorem 1.3. *The Riemann surface is connected.*

Proof. Let S be a connected component. Let $P \in S$ and let $z \in K$ be a function having a pole only at P (such a z exists by the Riemann-Roch theorem). Then z is holomorphic on any other component, without pole, hence constant, equal to c on such a component. But $z - c$ has infinitely many zeros, which is impossible since R and S are compact.

Theorem 1.4. *Let $z \in K$ be a non-constant function. The points of K induce points of $\mathbf{C}(z)$, and thereby induce a mapping of R onto the z-sphere S_z, which is a ramified topological covering. The algebraic ramification index e_P at a point P of R is the same as the topological index, and the number of sheets of the covering is*

$$n = [K : \mathbf{C}(z)].$$

Proof. Let P be a point of R and t a parameter at P. Then P induces a point $z = a$ on the z-sphere, and t^e is equal to $z - a$ times a unit in the power series ring $\mathbf{C}[[t]]$. Since one can extract nth roots in $\mathbf{C}$, we can find a local parameter u at P on R such that $u^e = z - a$, where $a = z(P)$. The map

$$u(Q) \mapsto u(Q)^e = z(Q) - a$$

gives an e to 1 map of a disc V_P around P onto a disc V_a in the z-plane. We shall call such a disc **regular** for P. We have shown that the topological ramification index is equal to e. As to the number of sheets, all but a finite number of primes of $\mathbf{C}(z)$ are unramified in K and, therefore, split completely into n primes of K (by the formula $\Sigma\, e_i = n$). Hence n is the number of sheets.

Theorem 1.5. *The Riemann surface is triangulable and orientable.*

Proof. We shall triangulate it in a special way, used later in another theorem. Let z be a non-constant function in K. Consider a triangulation of the z-sphere S_z such that every point q of S_z ramified in K is a vertex (just add such points to a given triangulation). A sufficiently high subdivision of the triangulation achieves the following properties.

If Δ is a triangle in S_z none of whose vertices is ramified, then Δ is contained in some regular neighborhood of a point, with ramification index equal to 1.

If Δ is a triangle with a ramified vertex Q, then $\Delta \subset V_Q$.

We can now life the triangulation. First, each vertex lifts to a certain number of vertices on R, and each Δ none of whose vertices is ramified lifts uniquely to n triangles in R.

If Q is a point of R ramified above q in S_z, we let m be its ramification index and t a parameter at Q on R. We get a map

$$V_Q \rightarrow V_q$$

by

$$t \mapsto t^m = z - a$$

if we choose t suitably. Each point $z - a = re^{i\theta}$ has m inverse images

$$t = r^{1/m} \exp\left(\frac{i\theta}{m} + \frac{2\pi i \nu\theta}{m}\right) \qquad \nu = 1, \ldots, m.$$

Hence each triangle Δ with a vertex at q lifts in m ways.

As to orientability, we can assume that two triangles of our triangulation, none of whose vertices are ramified and having an edge in common, are contained in some regular disc. Secondly, if Δ has a ramified vertex q, and Δ_1 has an edge in common with Δ, then Δ, Δ_1 are both contained in V_q.

Now we can orient the triangles on S_z such that an edge receives opposite orientation from the two triangles adjacent to it. If $\widetilde{\Delta}$ on R covers Δ, we give $\widetilde{\Delta}$ and its edges the same orientation as Δ. In view of our strengthened conditions, we can lift the orientation, so R is orientable.

Let V, E, T be the number of vertices, edges, triangles on the z-sphere. Then denoting by a prime the same objects on R, we get

$$E' = nE \qquad T' = nT.$$

On the other hand, let r be the number of primes p of $\mathbf{C}(z)$ which are ramified, or $p = \infty$. Let n_p be the number of primes P of K such that P lies above p. Then

$$V' = n(V - r) + \sum_p n_p,$$

the sum being taken over ramified p or infinity. We have

$$V' = nV - nr + \sum_P n_p$$

$$= nV + \sum_P (n_p - n)$$

$$= nV - \sum_P \sum_{P|p} (e_P - 1)$$

$$= nV - \sum_P (e_P - 1)$$

where this last sum is taken over all primes P of K.

Let X_z be the Euler characteristic of S_z and X_R that of R. We shall prove:

Theorem 1.6. *Let g be the topological genus of R. Then*

$$2g - 2 = -2n + \sum_P (e_P - 1).$$

Proof. We have $X_z = B_0 - B_1 + B_2$ where B_i is the i-th Betti number. We have $B_0 = B_2 = 1$. But also $X_z = V - E + T$, so $X_z = 2$.

Now $X_R = V' - E' + T'$ and by our previous result, this is

$$= nV - \sum (e_P - 1) - nE + nT$$

$$= nX_z - \sum (e_P - 1).$$

But $B_0' = B_2' = 1$ and $B_1' = 2g$. Hence $X_R = 2 - 2g$, whence

$$2g - 2 = -2n + \sum (e_P - 1)$$

as desired.

Corollary. *The algebraic genus is equal to the topological genus.*

Proof. Both satisfy the same formula.

§2. Integration on the Riemann Surface

As before, K is a function field over $\mathbf{C}$ and R its Riemann surface.

If U is an open subset of R, we can define the notion of **meromorphic** differential on U, namely an expression of type $f\,dx$ where f is a meromorphic function on U and x lies in K. We say that $f\,dx \sim g\,dy$ if $f/g = dy/dx$. We say that the differential is **holomorphic** at P if, whenever t is a parameter at P, and

$$f\,dx = f\frac{dx}{dt}\,dt$$

at P, then the function $f\,dx/dt$ has no pole at P. We say that the differential is holomorphic on U if it is so at every point of U.

A **primitive** g of $f\,dx$ on U is a meromorphic function on U such that $dg/dx = f$. If U is connected, then two primitives differ by a constant, as always.

If ω is a holomorphic differential on a disc V, then ω has a primitive g and g is holomorphic on V.

Let γ be a 1-simplex contained in a disc V, let $\omega = f\,dx$ on V, let

$$\partial\gamma = P - Q,$$

and let q be a primitive of ω on $\overline{V}$. Then we define

$$\int_\gamma \omega = g(P) - g(Q).$$

This number is independent of the choice of V and g. Indeed, if supp (γ) is contained in another disc V_1, and g_1 is a primitive of $f\,dx$ on V_1, then $V \cap V_1$ contains a connected open set W containing γ, and $g - g_1$ is constant on W.

If $\mathrm{subd}_1\,\gamma = \gamma_1 + \gamma_2$ has one more point A, then

$$\partial\gamma_1 = P - A,\ \partial\gamma_2 = A - Q \quad \text{and} \quad \int_\gamma \omega = \int_{\gamma_1} \omega + \int_{\gamma_2} \omega.$$

Hence if γ is contained in a disc V, then

$$\int_\gamma \omega = \int_{\mathrm{subd}_1\,\gamma} \omega.$$

If $\gamma = \Sigma\, n_i \sigma_i$ where σ_i is a 1-simplex contained in some open disc, we define the integral of ω over γ by linearity. We get again that the integral does not change by a subdivision.

If γ is any 1-chain, for r large enough, each member of the r-th subdivision is contained in some open disc, and we can define the integral over a 1-chain, independently of the subdivision chosen subject to this condition.

A cycle γ is called **homologous** to 0, and we write $\gamma \sim 0$, if some subdivision of γ is the boundary of a 2-chain. By abuse of language, we also say that γ itself is the boundary of a 2-chain.

Cauchy's Theorem. *Let ω be holomorphic on an open set U of R. Let γ be a 1-cycle on U, homologous to 0 on U. Then*

$$\int_\gamma \omega = 0.$$

Proof. We have $\gamma = \partial \eta$ where η is a 2-chain. For some r, $\text{Sd}^r \gamma$ has each one of its simplices contained in a disc. We have

$$\text{Sd}^r \gamma = \text{Sd}^r \partial \eta = \partial \text{Sd}^r \eta,$$

whence it suffices to prove Cauchy's theorem under the assumption that η and γ are contained in a disc. But then ω has a primitive, and the result is trivial.

Corollary. *If U is simply connected, then ω has a primitive on U.*

The pairing

$$(\gamma, \omega) \mapsto \int_\gamma \omega$$

gives a bilinear map

$$H_1(R, \mathbf{Z}) \times \Omega_1(R) \to \mathbf{C},$$

where Ω_1 is the space of differentials of first kind.

Theorem 2.1. *The kernels of this pairing on both sides are 0. In other words, if ω is a differential of first kind whose integral along every cycle is 0, then $\omega = 0$. Conversely, if γ is a cycle such that the integral along γ of every dfk is 0, then γ is homologous to 0.*

Proof. As to the first statement, fix a point O on R. Under the hypothesis that ω is orthogonal to every cycle, it follows that the association

$$P \mapsto \int_O^P \omega$$

is a holomorphic function on R, whence constant, and therefore that $\omega = 0$. The converse is harder to prove and will be done later, Theorem 5.4 of Chapter IV.

CHAPTER IV

The Theorem of Abel-Jacobi

§1. Abelian Integrals

A differential will be said to be of the **first kind** if it is holomorphic every-where on the Riemann surface. Such differentials form a vector space over the complex, and by the Riemann-Roch theorem, one sees that the dimension of this space is equal to the genus g of R.

From topology, we now take for granted that our Riemann surface can be represented as a polygon $\mathcal{P}$ with identified sides (Fig. 1).

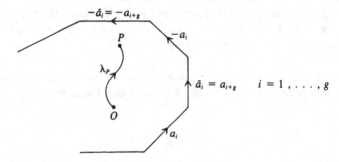

Figure 1

We select a point O inside the polygon, and use it as an origin. Let φ be a differential of first kind (written **dfk**). Given any point P inside the polygon (which is simply connected), we define

$$f(P) = \int_{\lambda_P} \varphi$$

where λ_P is any path from O to P lying entirely inside the polygon. Then f is single valued and holomorphic inside $\mathscr{P}$.

If λ_1 is an arbitrary path on R, not necessarily lying inside the polygon, from O to P, then

$$\lambda_1 \sim \lambda_P + \sum n_i a_i$$

with suitable integers (and $\sim$ means homologous). Consequently

$$\int_{\lambda_1} \varphi = \int_{\lambda_P} \varphi + \sum n_i \int_{a_i} \varphi.$$

The numbers

$$\alpha_i(\varphi) = \alpha_i = \int_{a_i} \varphi$$

generate an abelian group which will be called the group of **periods** of φ. An integral taken from O to P along any path is well defined modulo periods.

The integral over the path λ as shown in Fig. 2

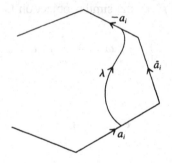

Figure 2

is also a period, and we write

$$\int_{\lambda} \varphi = \bar{\alpha}_i(\varphi).$$

Consider one of the cycles a_i. We can find a simply connected open set U containing a_i minus the vertex v. We can now define a holomorphic function $f_i^{\dagger}$ on U as follows. Given any point P on $a_i - v$, we take a path $\lambda_P^{\dagger}$ lying entirely inside the polygon except for its end point. For any point Q in U we then define

$$f_i^{\dagger}(Q) = \int_{\lambda_P^{\dagger}} \varphi + \int_P^Q \varphi,$$

the integral from P to Q being taken on a path lying entirely inside U. Our function f_i^+ is holomorphic on $a_i - v$.

Let ω be a differential which is holomorphic on a_i (including the vertex). We can define

$$\int_{a_i} f_i^+ \, \omega$$

as the limit of the integral

$$\int_{P_i}^{Q_i} f_i^+ \, \omega$$

as P_i, Q_i approach the vertex as indicated.

Similarly, we define f_i^- as the similar object on the side $-a_i$. We get

$$\int_{-a_i} f_i^- \, \omega = - \int_{a_i} f_i^- \, \omega.$$

Let

$$\widetilde{\alpha}_i(\varphi) = \int_{\widetilde{a}_i} \varphi.$$

Then for P on a_i we get

$$f_i^+(P) - f_i^-(P) = -\widetilde{\alpha}_i(\varphi).$$

We can now define the symbol

$$\int_{\mathcal{P}} f \omega$$

for any differential holomophic on the polygon $\mathcal{P}$ by

$$\int_{\mathcal{P}} f \omega = \sum_{i=1}^{2g} \int_{a_i} f_i^+ \, \omega + \sum \int_{-a_i} f_i^- \, \omega$$

$$= \sum_{i=1}^{2g} \lim_{\substack{P_i \to v \\ Q_i \to v}} \int_{P_i}^{Q_i} (f_i^+ - f_i^-) \, \omega$$

$$= \sum_{i=1}^{2g} \lim \int_{P_i}^{Q_i} - \tilde{\alpha}_i \omega$$

$$= - \sum_{i=1}^{2g} \tilde{\alpha}_i(\varphi) w_i$$

where

$$w_i = \int_{a_i} \omega$$

is a period of ω.

Theorem 1.1. *Let ω be a differential holomorphic on $\mathcal{P}$. Then*

$$\int_{\mathcal{P}} f\omega = - \sum_{i=1}^{2g} w_i \tilde{\alpha}_i = 2\pi\sqrt{-1} \sum \text{res}_P(f\omega)$$

where the sum over P is taken over poles of ω.

Proof. Our function f is uniquely defined inside the polygon. To get the integral over the polygon itself, where f has been defined separately for the two representations of the same side, we approximate the polygon $\mathcal{P}$ by a polygon $\mathcal{P}'$ as indicated (Fig. 3)

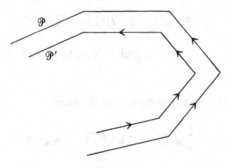

Figure 3

such that all the poles of ω lie inside $\mathcal{P}'$. Then

$$\lim_{\mathcal{P}' \to \mathcal{P}} \int_{\mathcal{P}'} f\omega = \int_{\mathcal{P}} f\omega,$$

the limit being taken as the vertices of $\mathcal{P}'$ approach the vertex of $\mathcal{P}$. The polygon $\mathcal{P}'$ is homologous to the sum of small circles taken with positive orientation around the poles of ω, and the integral around $\mathscr{P}'$ is, therefore, constant, equal to the expression on the right-hand side of our equality. Since it approaches the integral around $\mathscr{P}$, it is equal to it, thereby proving our theorem.

We formulate our preceding results more generally as follows:

Let $\Phi = (\varphi_1, \ldots, \varphi_g)$ be a basis for the dfk. We consider the vector integral

$$\int_{a_i} \Phi = \left(\int_{a_i} \varphi_1, \ldots, \int_{a_i} \varphi_g \right) = (\alpha_{i1}, \ldots, \alpha_{ig}) = A_i$$

and call it a **vector period**, or simply **period**, from now on. Similarly,

$$\int_{\widetilde{a}_i} \Phi = (\widetilde{\alpha}_{i1}, \ldots, \widetilde{\alpha}_{ig}) = \widetilde{A}_i.$$

Note that $\widetilde{A}_1, \ldots, \widetilde{A}_{2g}$ and $A_1, \ldots, A_{2g}$ differ by a permutation.

For P inside the polygon, we let

$$F(P) = \int_{\lambda_P} \Phi = (f_1(P), \ldots, f_g(P))$$

the f_ν corresponding to integrals of φ_ν. We define F_i^+ and F_i^- in a similar way, so that for P on a_i (and $\neq$ vertex) we have

$$F_i^+(P) - F_i^-(P) = -\widetilde{A}_i.$$

We can now define the vector integral $\int_{\mathscr{P}} F\omega$, and Theorem 1 can be formulated vectorially.

Theorem 1.2. *If ω is holomorphic on $\mathscr{P}$, then*

$$-\sum_{i=1}^{2g} w_i \widetilde{A}_i = 2\pi\sqrt{-1} \sum_P \operatorname{res}_P(F\omega),$$

the residue being the vector of residues, and the sum over P is taken over all poles of ω.

§2. Abel's Theorem

We have fixed a point O on the Riemann surface. The vector integral

$$\int_O^P \Phi = \left(\int_O^P \varphi_1, \ldots, \int_O^P \varphi_g \right)$$

as P varies over the Riemann surface, taken along some path from O to P, is well-defined modulo periods. In this way, we obtain a mapping from the Riemann surface R into the factor group $\mathbf{C}^g$ modulo periods. This mapping can be extended by linearity to the free abelian group generated by points, this group being called the group of **cycles** on R, or **divisors**. A divisor is a formal sum

$$\mathfrak{a} = \sum n_P P$$

with integers n_P, almost all of which are 0. Its **degree** is the sum of the n_P. Those of degree 0 form a group $\mathfrak{D}_0$, and for these it is clear that the restriction of the above homomorphism is then independent of the origin chosen. We thus obtain a homomorphism

$$\mathfrak{a} \mapsto S(\mathfrak{a}) \equiv \sum n_P \int_O^P \Phi \quad \text{mod periods,}$$

from $\mathfrak{D}_0$ into $\mathbf{C}^g/(\text{periods})$. This factor group will be denoted by J and is called the **Jacobian** of R. The main theorem is the following statement.

Abel-Jacobi Theorem. *The preceding homomorphism from $\mathfrak{D}_0$ into J is surjective, and its kernel consists of the divisors of functions. It establishes an isomorphism between the divisor classes (for linear equivalence) of degree 0, and the Jacobian group $\mathbf{C}^g$ modulo periods.*

The statement concerning the kernel of our homomorphism is called **Abel's theorem** and will be proved in this section. The surjectivity is postponed to the next seciton.

We first prove that divisors of functions are contained in the kernel. In other words, if z is a function with divisor

$$(z) = \sum n_P P,$$

we have to show that

$$\sum n_P F(P) \equiv 0 \quad \text{mod periods.}$$

We can always find a polygon representation of R such that z has no pole on $\mathscr{P}$, and we let $\omega = dz/z$ in Theorem 1.2. Then $\text{res}_P(F\omega) = n_P F(P)$. Indeed, for any holomorphic function f consider a local parameter t at a point P. Then $f\omega = f(dz/dt)z^{-1}\,dt$. Also,

$$z^{-1} \frac{dz}{dt} = \frac{k}{t} + \cdots$$

if z has order k at P. Thus

$$\text{res}_t \left(f z^{-1} \frac{dz}{dt} \right) = f(0) k.$$

By Theorem 1.2,

$$2\pi \sqrt{-1} \sum n_P F(P) = - \sum w_i \widetilde{A}_i$$

where $w_i = \int_{a_i} z^{-1} \, dz$. But it is easily shown that $w_i = 2\pi \sqrt{-1} \, m_i$ for some integer m_i. Hence we cancel $2\pi \sqrt{-1}$ and we get

$$\sum n_P F(P) = - \sum_{i=1}^{2g} m_i \widetilde{A}_i$$

as was to be shown.

For the convenience of the reader, we recall the proof that if z is a function which is holomorphic on a cycle a, then

$$\int_a \frac{dz}{z} = \text{integral multiple of } 2\pi \sqrt{-1}.$$

We may assume that a is a closed path. The integral is defined by analytic continuation along the path, over successive discs, say $D_0, \ldots, D_N$ starting from a point P_0 and returning to P_0. Let L_k be a primitive of dz/z over the disc D_k. Let P_k be a point in $D_{k+1} \cap D_k$ so that

$$\int_a \frac{dz}{z} = L_0(P_1) - L_0(P_0) + L_1(P_2) - L_1(P_1) + \cdots.$$

All terms cancel except $L_N(P_N) - L_0(P_0)$. But L_N and L_0 are two primitives for dz/z over a disc around P_0, and so they differ by an integral multiple of $2\pi \sqrt{-1}$. Since $P_N = P_0$, this proves what we wanted.

In order to prove the converse, we need some lemmas.

Lemma 2.1. *Let x_i $(i = 1, \ldots, 2g)$ be complex numbers. Then*

$$\sum x_i \widetilde{A}_i = 0 \quad \text{if and only if} \quad x_i = B \cdot A_i$$

for some complex vector $B = (b_1, \ldots, b_g)$. The vectors $A_1, \ldots, A_{2g}$ span a g-dimensional space over the complex numbers.

Proof. We prove first that the relation $X \cdot A_i = 0$ for all i implies that $X = 0$ (for a complex vector X). Let $\omega = X \cdot \Phi$. Then

$$\int_{a_i} \omega = \int_{a_i} X \cdot \Phi = X \cdot \int_{a_i} \Phi = X \cdot A_i = 0$$

by hypothesis. Hence all the periods of ω are 0. Hence the integral $\int_0^P \omega$ is a holomorphic function on R without poles because ω is of first kind. It is therefore, constant; hence $\omega = 0$ and $X = 0$ because the φ_ν are linearly independent over $\mathbf{C}$.

This shows that $A_1, \ldots, A_{2g}$ generate $\mathbf{C}^g$.

Let $\omega = b_1 \varphi_1 + \cdots + b_g \varphi_g$ be a dfk. Then $F\omega$ has no poles and hence by Theorem 1.2,

$$0 = - \sum \operatorname{res}_P (F\omega) = \frac{-1}{2\pi\sqrt{-1}} \sum w_i \widetilde{A}_i$$

where

$$w_i = \sum_{\nu=1}^g b_\nu \int_{a_i} \varphi_\nu = \sum b_\nu a_{i\nu} = B \cdot A_i.$$

This proves half the first assertion.

Conversely, we want to find the space of relations

$$x_1 \widetilde{A}_1 + \cdots + x_{2g} \widetilde{A}_{2g} = 0.$$

We know from the fact that $A_1, \ldots, A_{2g}$ generate $\mathbf{C}^g$ that the relations

$$X = (B \cdot A_1, \ldots, B \cdot A_{2g})$$

form a g-dimensional space, which must therefore be the whole space of all relations. This proves the lemma.

A differential ω is said to be of **second kind** if its residue is 0 at all points. It is said to be of **third kind** if it has at most poles of order 1 at all points. The next lemma concerns such differentials, abbreviated by **dtk**.

Lemma 2.2. *Let $\mathfrak{a} = \Sigma \, n_P P$ be a divisor of degree 0. Then there exists a differential of third kind ω such that $\operatorname{res}_P \omega = n_P$ for all P.*

Proof. It suffices to prove our assertion for the case where all coefficients are zero except for two points. Indeed, by induction, suppose it proved for points $P_1, \ldots, P_r$. Let Q be a new point unequal to the given points. Let ω_j ($j = 1, \ldots, r$) have poles only at P_j and Q such that $\mathrm{res}_{P_j} \omega_j = n_j$ and $\mathrm{res}_Q \omega_j = -n_j$. We let $\omega = \Sigma \, \omega_j$. It obviously has the required property.

The Riemann-Roch theorem gives

$$l(-\mathfrak{a}) = -\deg(\mathfrak{a}) + 1 - g + d(\mathfrak{a})$$

where $d(\mathfrak{a})$ is the dimension of the space of differentials ω such that

$$(\omega) \geqq -\mathfrak{a}.$$

Put $\mathfrak{a} = P_1 + P_2$. We get $d(\mathfrak{a}) = g + 1$ and hence there exists a dtk ω having a pole of order 1 at P_1 or P_2. But the sum of the residues is 0, so it actually has a pole at both, with opposite residues. Multiplying ω by a suitable constant gives what we want.

We can now determine the kernel of our map. Let $\mathfrak{a} = \Sigma \, n_P P$ be a divisor of degree 0 such that $S(\mathfrak{a}) = 0$. We have to show that $\mathfrak{a}$ is the divisor of a function.

We contend that there exists a dtk ψ such that ψ has poles at P with residues n_P, and such that

$$\int_{a_i} \psi = 2\pi \sqrt{-1} \, n_i$$

for suitable integers n_i. By the lemma, there exists a dtk ω having residue n_P at P for all P. For any P, we have

$$\mathrm{res}_P (F\omega) = n_P F(P).$$

We are going to change ω by a dfk, so as to change the periods but not the residues.

By Theorem 1.2 we get

$$-\sum w_i \tilde{A}_i = 2\pi\sqrt{-1} \sum \mathrm{res}_P (F\omega)$$

$$= 2\pi\sqrt{-1} \sum n_P F(P)$$

$$= 2\pi\sqrt{-1} \sum m_i \tilde{A}_i$$

for some integers m_i, by hypothesis. Hence

$$\sum (w_i - 2\pi\sqrt{-1}\, m_i)\widetilde{A}_i = 0.$$

By Lemma 2.1, we have

$$w_i - 2\pi\sqrt{-1}\, m_i = B \cdot A_i$$

for some vector B. We let $\psi = \omega - B \cdot \Phi$. Then ψ has the same poles and residues as ω. Furthermore,

$$\int_{a_i} \psi = \int_{a_i} \omega - \int_{a_i} B \cdot \Phi = w_i - B \cdot A_i = 2\pi\sqrt{-1}\, m_i,$$

thereby proving our contention.

To construct the desired function z we take essentially

$$z(P) = \exp \int_O^P \psi,$$

O being as usual a suitable origin inside our polygon. For any point P which is not a pole of ψ, and any path from O to P, the exponential of the integral gives a function which is independent of the path and is thus a well-defined meromorphic function. For a pole of ψ, expanding ψ in terms of a local parameter around this point shows at once that we get a meromorphic function around the point, whose singularity can only be a pole. Abel's theorem is proved.

§3. Jacobi's Theorem

We must now prove that our map from divisors of degree 0 into the factor group $\mathbf{C}^g$ modulo periods is surjective.

A divisor $\mathfrak{a}$ is said to be **non-special** if $d(-\mathfrak{a}) = 0$, that is if there exists no differential ω such that $(\omega) \geqq \mathfrak{a}$.

Lemma 3.1. *There exists g distinct points $M_1, \ldots, M_g$ such that the divisor $M_1 + \cdots + M_g$ is non-special.*

Proof. Let $\omega_1 \neq 0$ be a dfk, and M_1 a point which is not zero of ω_1. The space of dfk having a zero at M_1 has dimension $g - 1$ by Riemann-Roch. Let $\omega_2 \neq 0$ be in it and let M_2 be a point which is not a zero of ω_2. Continue g times to get g points, as desired.

Theorem 3.2. *Let $M_1, \ldots, M_g$ be g distinct points such that the divisor $M_1 + \cdots + M_g$ is non-special. Then the map*

$$(P_1, \ldots, P_g) \mapsto \sum \int_{M_i}^{P_i} \Phi$$

gives an analytic isomorphism of a product of small discs $V_1 \times \cdots \times V_g$ around the points $M_1, \ldots, M_g$ onto a neighborhood of zero in $\mathbb{C}^g$.

Before proving our theorem, we show how it implies Jacobi's theorem.

First we note that it suffices to prove Jacobi's theorem for a neighborhood of zero in $\mathbb{C}^g$, that is, to prove local surjectivity. Indeed, let X be any vector. For large n, $1/n \cdot X$ is in a small neighborhood of zero so, by the local result, we can find a divisor of degree zero such that $F(\alpha) = 1/n \cdot X$ modulo periods. Then $F(n\alpha) \equiv X$ modulo periods.

Using the notation of Theorem 3.2, let α range over divisors

$$P_1 + \cdots + P_g - (M_1 + \cdots + M_g),$$

with P_i ranging over V_i. Then clearly

$$F(\alpha) \equiv \sum \int_{M_i}^{P_i} \Phi \qquad \text{(mod periods)}$$

and the theorem implies what we want.

We now prove Theorem 3.2. Let t_i be a local parameter at M_i. We contend that the determinant

$$\begin{vmatrix} \dfrac{\varphi_1}{dt_1}(M_1) & \cdots & \dfrac{\varphi_1}{dt_g}(M_g) \\ & \cdots & \\ \dfrac{\varphi_g}{dt_1}(M_1) & \cdots & \dfrac{\varphi_g}{dt_g}(M_g) \end{vmatrix}$$

is not zero. To prove this, consider the homomorphism

$$\varphi \mapsto \left[\frac{\varphi}{dt_1}(M_1), \ldots, \frac{\varphi}{dt_g}(M_g) \right]$$

of differentials of first kind into complex g-space. Any dfk in the kernel would be special, so the kernel is trivial, and our map is a linear isomorphism. Hence our determinant is non-zero.

We write

$$\varphi_1 = h_{11}(t_1)dt_1, \ldots, \varphi_1 = h_{1g}(t_g)dt_g$$
$$\cdots$$
$$\varphi_g = h_{g1}(t_1)dt_1, \ldots, \varphi_g = h_{gg}(t_g)dt_g$$

where $h_{ij}(t_j)$ is holomorphic in V_j. Let $h_{ij}^*(t_j)$ be the integral of the power series $h_{ij}(t_j)$, normalized to vanish at $t_j = 0$. Then we get a representation of our map:

$$(P_1, \ldots, P_g) \mapsto (h_{11}^*(t_1) + \cdots + h_{1g}^*(t_g), \ldots)$$

$$= (H_1(t_1, \ldots, t_g), \ldots, H_g(t_1, \ldots, t_g))$$

where each $H_i(t_1, \ldots, t_g)$ is holomorphic in the g variables, on

$$V_1 \times \cdots \times V_g.$$

Now

$$\frac{\partial H_i}{\partial t_j} = h_{ij}(t_j),$$

and hence the Jacobian determinant evaluated at $(0, \ldots, 0)$ is precisely our preceding determinant, and is non-zero. By the implicit function theorem, we get our local analytic isomorphism, thereby proving Jacobi's theorem.

The theorem can be complemented by an important remark. We consider complex g-space as a real $2g$-dimensional space. We have:

Theorem 3.3. *The periods $A_1, \ldots, A_{2g}$ are linearly independent over the reals. Hence the factor space of $\mathbf{C}^g$ by the periods is a $2g$-real-dimensional torus.*

Proof. It will suffice to prove that any complex vector X is congruent to a vector of bounded length modulo periods. By the Riemann-Roch theorem, any divisor of degree 0 is linearly equivalent to a divisor of type

$$(P_1 + \cdots + P_g) - g \cdot O,$$

for suitable points $P_1, \ldots, P_g$, which may be viewed as lying inside or on the polygon. For P ranging over the Riemann surface, the integrals

$$\int_O^P \Phi$$

are bounded in the Euclidean norm, if we take the path of integration to be entirely inside the polygon, except if P lies on the boundary. Consequently, the sum

$$\sum_{i=1}^{} \int_O^{P_i} \Phi$$

has also bounded norm (g times the other bound). Combining this with Jacobi's surjectivity theorem shows that any vector is congruent to one with bounded length modulo periods, and concludes the proof of Theorem 3.3.

§4. Riemann's Relations

Let φ, φ' be two dfk, and consider the polygon representation of R, with sides a_i, b_i, $-a_i$, $-b_i$ ($i = 1, \ldots, g$) following each other (Fig. 4). We let

$$\alpha_i = \int_{a_i} \varphi \quad \text{and} \quad \beta_i = \int_{b_i} \varphi$$

be the canonical periods of φ, and α'_i, β'_i those of φ'.

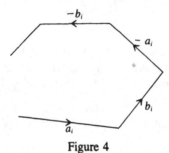

Figure 4

Riemann's first relations *state that*

$$\sum_{i=1}^{g} (\alpha_i \beta'_i - \alpha'_i \beta_i) = 0.$$

Proof. Let $f(P)$ be the function defined before (integral from O to P, suitably defined on the boundary of the polygon), and consider the integral

$$\int_{\mathscr{P}} f \varphi'.$$

Since φ' has no poles, we get from Theorem 1.2

$$0 = \sum_{i=1}^{g} \left(\int_{a_i} \varphi' \int_{b_i} \varphi + \int_{b_i} \varphi' \int_{-a_i} \varphi \right)$$

which gives what we want.

Riemann's second relations *state that*

$$\frac{1}{2\sqrt{-1}} \sum_{j=1}^{g} (\bar{\alpha}_j \beta_j - \bar{\beta}_j \alpha_j) > 0.$$

Proof. This time, we consider the integral

$$\frac{1}{2i} \int \bar{f}\, \varphi = \text{left hand side of inequality,}$$

and we prove that it is strictly positive. Write

$$f = u + iv,$$

in terms of its real and imaginary part. Then $\bar{f} = u - iv$, and

$$df = \varphi = du + i\, dv.$$

Then

$$\bar{f}\varphi = \tfrac{1}{2}d(u^2 + v^2) + i(u\, dv - v\, du).$$

The first term on the right is exact, and the second gives

$$d(u\, dv - v\, du) = 2du \wedge dv.$$

By Green's theorem, our integral can be replaced by the double integral

$$2i \iint du \wedge dv.$$

In the neighborhood of each point, we may take a chart, and express the functions u, v in terms of a complex variable z, with real parts x, y, that is $z = x + iy$. By the Cauchy-Riemann equations, we get

$$du \wedge dv = \left(\left(\frac{\partial u}{\partial x}\right)^2 + \left(\frac{\partial u}{\partial y}\right)^2 \right) dx \wedge dy,$$

which shows that the integral is positive, as desired.

§5. Duality

Lemma 5.1. *Numbering the sides of the polygon as at the beginning, $a_1, \ldots, a_{2g}$, and given an index i, there exists a dfk φ with periods satisfying*

$$\text{Re} \int_{a_i} \varphi = 1, \qquad \text{Re} \int_{a_j} \varphi = 0 \qquad \text{for} \qquad j \neq i.$$

Proof. Consider the period matrix:

	a_1	$\cdots$	a_{2g}
φ_1	$u_1^1 + iv_1^1$	$\cdots$	$u_{2g}^1 + iv_{2g}^1$
$\vdots$	$\vdots$		$\vdots$
φ_g	$u_1^g + iv_1^g$	$\cdots$	$u_{2g}^g + iv_{2g}^g$

where the u, v are the real and imaginary parts of the periods, and are, therefore, real. Then the column vectors of the above matrix are our periods $A_1, \ldots, A_{2g}$. We wish to find g complex numbers

$$z_k = x_k + iy_k \ (k = 1, \ldots, g) \quad \text{and a dfk} \quad \varphi = z_1\varphi_1 + \cdots + z_g\varphi_g$$

having the required periods. This amounts to solving, say for $i = 1$, the system of equations

$$x_1 u_1^1 - y_1 v_1^1 + \cdots + x_g u_1^g - y_g v_1^g = 1$$
$$x_1 u_2^1 - y_1 v_2^1 + \cdots + x_g u_2^g - y_g v_2^g = 0$$
$$\cdots$$

This is solvable if the row vectors of the coefficient matrix are linearly independent over the reals. This is indeed the case, because a linear relation is immediately seen to imply a linear relation between $A_1, \ldots, A_{2g}$ over the reals.

Lemma 5.2. A dfk *cannot have all its periods pure imaginary.*

Proof. The preceding system of linear equations cannot have a solution when made homogeneous.

Theorem 5.3. *Given a divisor* $\mathfrak{a} = \Sigma \, n_P P$ *of degree* 0, *there exists a unique dtk* $\omega_{\mathfrak{a}}$ *such that*

(1) $\text{res}_P \, \omega_{\mathfrak{a}} = n_P$ *for all P.*

(2) *The periods of* $\omega_{\mathfrak{a}}$ *are all pure imaginary.*

Proof. Let ω be a dtk with $\text{res}_P \, \omega = n_P$. It suffices to find a dfk φ having the same real parts for the canonical periods (integrals around a_i). This is immediate from Lemma 5.1. The uniqueness is clear, because the difference between two differentials satisfying the conditions of the theorem would be a dfk contradicting Lemma 5.2.

Theorem 5.4. *If σ is a cycle such that $\int_\sigma \varphi = 0$ for all dfk then $\sigma \sim 0$.*

Proof. Let $\sigma \sim n_1 a_1 + \cdots + n_{2g} a_{2g}$. Find a dfk φ having periods such that

$$\text{Re} \int_{a_1} \varphi = 1, \quad \text{and} \quad \text{Re} \int_{a_j} \varphi = 0, \quad j \neq 1.$$

Then

$$0 = \int_\sigma \varphi = n_1 + \text{pure imaginary},$$

and so $n_1 = 0$. Similarly, $n_j = 0$ all j.

Theorem 5.5. *The pairing*

$$(\sigma, \alpha) \mapsto \langle \sigma, \alpha \rangle = \exp \int_\sigma \omega_\alpha$$

induces a bilinear pairing between the first homology group $H_1(R)$ and the group of divisor classes of degree 0. Its kernels on both sides are 0, and the pairing induces the Pontrjagin duality between the discrete group $H_1(R)$ and the compact torus.

Proof. Let $\mathcal{D}_0$ be the group of divisors of degree 0. If $\alpha \in \mathcal{D}_0$ and $\alpha = (z)$ is the divisor of a function, then we may take $\omega_\alpha = dz/z$. All the periods of dz/z are pure imaginary, of the form $2\pi\sqrt{-1}\, m$ with some integer m. Hence $\langle \sigma, \alpha \rangle = 1$ for such divisors, and all cycles σ having no point in common with α. Furthermore, if $\sigma \sim 0$, then

$$\int_\sigma \omega_\alpha = 2\pi\sqrt{-1} \sum \text{res}_P\, \omega_\alpha$$

$$= 2\pi\sqrt{-1} \quad \text{times an integer},$$

so $\langle \sigma, \alpha \rangle = 1$.

Our pairing is therefore well defined as a bilinear map of $H_1(R) \times J$ into the circle.

If α is orthogonal to $H_1(R)$, then by definition, all periods of ω_α are of the type $2\pi\sqrt{-1}\, m$ for some integer m. We can then define a function z by letting

$$z(P) = \exp \int_O^P \omega_\alpha.$$

This function is meromorphic and its divisor is α. Hence the kernel on the right consists precisely of the group of divisors of functions.

To show that the kernel on the left consists precisely of these cycles σ which are ~ 0, let us take an arbitrary cycle $\sigma \sim \Sigma \, n_i a_i$. We must show that all $n_i = 0$. Let $M_1, \ldots, M_g$ be non-special points as in the proof of Jacobi's theorem, and let us assume without loss of generality that these points do not lie on the cycles a_i. We consider divisors of degree 0 of type

$$\alpha = P_1 + \cdots + P_g - M_1 - \cdots - M_g$$

where $P_1, \ldots, P_g$ lie in small neighborhoods $V_1, \ldots, V_g$ of $M_1, \ldots, M_g$ respectively. Thus

$$(P_1, \ldots, P_g) \in V_1 \times \cdots \times V_g \subset \mathbf{C}^g.$$

We put

$$w_i(\alpha) = \int_{a_i} \omega_\alpha$$

so that by Theorem 1.2,

$$w_1(\alpha)\tilde{A}_1 + \cdots + w_{2g}(\alpha)\tilde{A}_{2g} = -2\pi\sqrt{-1} \sum_{\nu=1}^{g} (F(P_\nu) - F(M_\nu))$$

$$= -2\pi\sqrt{-1} \sum_{\nu=1}^{g} \int_{M_\nu}^{P_\nu} \Phi.$$

If $\langle \sigma, \alpha \rangle = 0$ for all α, then

$$\sum n_i \int_{a_i} \omega_\alpha = \sum n_i w_i(\alpha) \in \mathbf{Z} \cdot 2\pi\sqrt{-1}.$$

Hence the points

$$(w_1(\alpha), \ldots, w_{2g}(\alpha)) \in i\mathbf{R}^{2g}$$

lie in a denumerable union of hyperplanes defined by equations

$$\sum_{i=1}^{2g} n_i v_i = m \cdot 2\pi\sqrt{-1}, \qquad m \in \mathbf{Z}.$$

However these points form an open set in $\sqrt{-1}\mathbf{R}^{2g}$, as one sees from the proof of Jacobi's theorem giving a local analytic isomorphism of

$$V_1 \times \cdots \times V_g.$$

This is impossible unless $n_i = 0$ for all i, because a hyperplane has measure 0, and concludes the proof of Theorem 5.5.

We may also return to the duality of Chapter III, Theorem 2.1. Theorem 5.4 proved the expected converse which we had to postpone before, but more is true.

Theorem 5.6. *The bilinear map*

$$H_1(R, \mathbf{Z}) \times \Omega_1(R) \to \mathbf{R}$$

given by

$$(\gamma, \omega) \mapsto \mathrm{Re} \int_\gamma \omega$$

induces a duality of real vector spaces, making $H_1(R, \mathbf{R})$ the $\mathbf{R}$-dual of the space of dfk.

Proof. By Lemma 5.2, the kernel of the pairing on the right, i.e. in the space of dfk, is equal to 0. As a real vector space, the dfk have dimension $2g$, by Lemma 5.1. Since $H_1(R, \mathbf{Z})$ has real dimension $\leq 2g$, it follows that its image in the dual space of $\Omega_1(R)$ has to generate a space of real dimension exactly $2g$, and the spaces are in duality with each other.

CHAPTER V

Periods on the Fermat Curve

We return to the special case of the Fermat curve, and compute the period lattice explicitly in terms of the basis for the differentials of first kind given in Chapter II.

The first two sections deal with general results on the universal covering space, concerning certain integrals of third kind, of type $d\log f$ where f is a nowhere vanishing function. These are then applied to the Fermat curve, and we follow Rohrlich [Ro 1], [Ro 2].

On the other hand, as a preliminary to these more exact descriptions of the period lattice, we can make a simpler analysis of the space generated by this lattice over the rationals, as follows. Let

$$\gamma_0 : [0, 1] \to F(N)$$

be the curve such that

$$t \mapsto (t, (1 - t^N)^{1/N}),$$

where the N-th root is the real N-th root. Then

$$\int_{\gamma_0} x^{r-1} y^{s-1} \frac{dx}{y^{N-1}} = \int_0^1 t^{r-1}(1 - t^N)^{(s-1)/N} \frac{dt}{(1 - t^N)^{(N-1)/N}} \cdot$$

Making the change of variables $u = t^N$, $du = Nt^{N-1}dt$, we find at once that

$$\int_{\gamma_0} \omega_{r,s} = \frac{1}{N} B\left(\frac{r}{N}, \frac{s}{N}\right),$$

where B is the Euler Beta function, given by

$$B(a, b) = \int_0^1 u^{a-1}(1 - u)^{b-1}\, du.$$

Let $\zeta = e^{2\pi i/N}$. We may take the images of the curve γ_0 under the automorphisms of $F(N)$ as indicated below. Then the chain

$$\kappa = \gamma_0 - (1, \zeta)\gamma_0 + (\zeta, \zeta)\gamma_0 - (\zeta, 1)\gamma_0$$

is in fact a closed path, i.e. a cycle on the Fermat curve. By the change of variables formula, one then finds at once

$$\int_\kappa \omega_{r,s} = (1 - \zeta^s + \zeta^{r+s} - \zeta^r)\frac{1}{N} B\left(\frac{r}{N}, \frac{s}{N}\right)$$

$$= (1 - \zeta^r)(1 - \zeta^s)\frac{1}{N} B\left(\frac{r}{N}, \frac{s}{N}\right).$$

This is a period of the differential form $\omega_{r,s}$, and it is $\neq 0$ if r, $s \not\equiv 0 \bmod N$. From this we shall prove:

$H_1(F(N), \mathbf{Q})$ *is generated by* κ *over* $\mathbf{Q}[G]$, *where* G *is the group of automorphisms of* $F(N)$ *which is naturally* $\approx \mu_N \times \mu_N$.

Indeed, for each character X of $\mu_N \times \mu_N$, the idempotent e_χ lies in the group ring $\mathbf{Q}[G]$, and for every differential of first kind $\omega_{r,s}$ we have

$$\int_{e_\chi \kappa} \omega_{r,s} = \int_\kappa e_\chi^* \omega_{r,s}.$$

Since $\omega_{r,s}$ is an eigenvector for the operation of the group algebra with character $X_{r,s}$, it follows from the period computation above that if ω is a differential form such that

$$\int_{e_\chi \kappa} \omega = 0 \qquad \text{for all } X$$

then $\omega = 0$. In other words, if ω is orthogonal to the module generated over $\mathbf{Q}[G]$ by κ, then $\omega = 0$. Our assertion follows from Theorem 5.6 of Chapter IV.

In the next sections, we develop the tools necessary to get the structure of the period module over $\mathbf{Z}$.

§1. The Logarithm Symbol

Let R be the Riemann surface associated with the function field K, and let $S = S_\infty$ be a finite non-empty set of points, which we shall call the points at **infinity**. If $f \in K$ has zeros and poles only in S, then we say that f is a **S-unit**, or simply a **unit** if S is fixed throughout. We denote the group of S-units by $E(S)$, or simply E.

Let $R' = R - S$ be the surface obtained by deleting the points of S from R, and let

$$p: U \to R'$$

be the universal covering space of R'. If $f \in E$ then $f \circ p$ is holomorphic non-zero on U, so there exists a determination of $\log f \circ p$ such that

$$\exp \log f \circ p = f \circ p.$$

Let $G = \text{Aut}(U, p)$ be the group of covering automorphisms, and let $\sigma \in G$. The function

$$z \mapsto (\log f \circ p)(\sigma z) - (\log f \circ p)(z)$$

is constant on U. Indeed,

$$\exp \log f \circ p(\sigma z) = f \circ p(\sigma z) = f \circ p(z),$$

so $(\log f \circ p) \circ \sigma$ is another choice of a branch of the logarithm, so it differs by a constant since U is connected.

For any $z \in U$, we have

$$\log f \circ p(\sigma z) - \log f \circ p(z) = \int_z^{\sigma z} d\log f \circ p = \int_\gamma \frac{df}{f},$$

where the integral from z to σz is taken along any curve (independent of the curve since U is simply connected), and the integral over γ is taken on the projection of that curve into the Rieman surface. Therefore there exists an integer $L(f, \sigma)$ such that for all z,

$$\log f \circ p(\sigma z) - \log f \circ p(z) = 2\pi i L(f, \sigma).$$

Lemma 1.1. *The symbol $L(f, \sigma)$ is bimultiplicative in f and σ.*

Proof. The formulas

$$L(fg, \sigma) = L(f, \sigma) + L(g, \sigma) \quad \text{and} \quad L(f, \sigma\sigma') = L(f, \sigma) + L(f, \sigma')$$

are trivially proved from the definitions, and the fact that the integral of $d\log f \circ p$ between two points of U does not depend on the choice of paths between the two points.

Thus we get a pairing

$$E \times \mathrm{Aut}(U, p)^{\mathrm{ab}} \to \mathbf{Z},$$

where $\mathrm{Aut}(U, p)^{\mathrm{ab}} = G^{\mathrm{ab}}$ is the factor group of G by its commutator group, namely

$$G^{\mathrm{ab}} = G/G^c.$$

Let m be a positive integer, and let $f \in E$ be a unit. Then $f \circ p$ has an m-th root on U, given by

$$\exp \frac{1}{m} \log(f \circ p).$$

If φ is any m-th root (any two differ by an m-th root of unity), then for any $\sigma \in G$ we get

$$\varphi^{\sigma} \varphi^{-1} = \exp \frac{2\pi i}{m} L(f, \sigma) = \zeta^{L(f,\sigma)}$$

where $\zeta = e^{2\pi i/m}$, and hence

$$\varphi^{\sigma} = \zeta^{L(f,\sigma)} \varphi.$$

§2. Periods on the Universal Covering Space

Let $\Omega = (\omega_1, \ldots, \omega_g)$ be a basis for the dfk on R. Again let

$$p: U \to R'$$

be the universal covering space as in the preceding section. For each

$$\sigma \in G = \mathrm{Aut}(U, p) \quad \text{and} \quad u \in U$$

let l_σ be a path in U from u to σu. Then the homology class of $p \circ l_\sigma$ in $H_1(R', \mathbf{Z})$ is independent of the choice of u. Indeed, if z is another point, then we have a "quadrilateral" in U as shown on the figure.

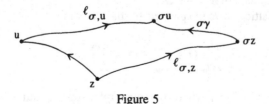

Figure 5

Its projection in R' looks like:

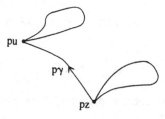

Figure 6

We have for any dfk ω:

$$\int_{p \circ l_{\sigma,u}} \omega = \int_{l_{\sigma,u}} p^*\omega = \int_u^{\sigma u} p^*\omega.$$

But the integral around the quadrilateral of $p^*\omega$ is equal to zero, because U is simply connected. Furthermore,

$$\int_{\sigma z}^{\sigma u} p^*\omega = \int_{p\sigma z}^{p\sigma u} \omega = \int_{pu}^{pz} \omega,$$

so the independence of the homology class of $p \circ l_\sigma$ follows by duality between cycles and differential forms.

The association

$$\sigma \mapsto \text{class of } p \circ l_\sigma \text{ in } H_1(R', \mathbf{Z})$$

gives an isomorphism

$$G/G^c \approx H_1(R', \mathbf{Z}),$$

where G^c is the commutator subgroup. Since we have a surjective homomorphism

$$H_1(R', \mathbf{Z}) \to H_1(R, \mathbf{Z}) \to 0,$$

induced by the inclusion $R' \subset R$, we can represent an element of $H_1(R, \mathbf{Z})$ by a cycle $p \circ l_\sigma$ for some $\sigma \in G$.

The period lattice of R is spanned by the vectors

$$\int_{p \circ l_\sigma} \Omega = \int_{l_\sigma} p^*\Omega = \int_u^{\sigma u} p^*\Omega$$

for any choice of $u \in U$, where σ ranges over representatives of G/G^c.

§3. Periods on the Fermat Curve

We apply the considerations of the first two sections to the Fermat curve $F(N)$ defined by

$$x^N + y^N = 1.$$

Note that $F(1)$ is isomorphic to projective 1-space $\mathbf{P}^1$, by the function x. If we delete the points $\{0, 1\}$, we obtain an affine curve (of genus 0). We let

$$\lambda : U \to \mathbf{C} - \{0, 1\}$$

be the universal covering space. (The choice of symbol λ is not accidental. It arises because it is the classical symbol for the classical function generating the modular function field of level 2–see later in the book.) We let

$$\Phi = \mathrm{Aut}\,(U, \lambda).$$

We take the set of points at infinity on $F(N)$ to be the inverse image of 0, 1 on $F(1)$. This is therefore the set of points

$$(0, \zeta^j, 1), \qquad (\zeta^j, 0, 1), \qquad (\zeta^j, \epsilon\zeta^j, 0),$$

where $\zeta = e^{2\pi i/N}$ and $\epsilon = e^{\pi i/N}$. The above coordinates are the projective coordinates (x, y, z) for the projective equation

$$x^N + y^N = z^N.$$

As before, we let $F(N)'$ be the open affine curve obtained by deleting the points of infinity as above from $F(N)$. It corresponds to an intermediate covering between U and $F(1)'$, and we let

$$\Phi(N) = \mathrm{Aut}\,(U/F(N)').$$

The relevant maps are shown in the following diagram.

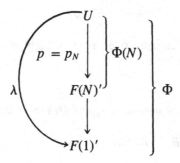

We let

$$\tilde{x} = x \circ p \qquad \text{and} \qquad \tilde{y} = y \circ p,$$

so $\tilde{x}$ and $\tilde{y}$ are N-th roots of λ and $1 - \lambda$ respectively on U. In the present case, let us abbreviate

$$L(\lambda, \sigma) = \alpha(\sigma) \qquad \text{and} \qquad L(1 - \lambda, \sigma) = \beta(\sigma).$$

Then by §1, for $\sigma \in \Phi$ we have

$$\tilde{x} \circ \sigma = \zeta^{\alpha(\sigma)} \qquad \text{and} \qquad \tilde{y} \circ \sigma = \zeta^{\beta(\sigma)} y.$$

From these relations, we see that

$$\Phi(N) = \text{set of } \sigma \in \Phi \text{ such that } \alpha(\sigma) \equiv \beta(\sigma) \equiv 0 \ (\text{mod } N).$$

We take for granted from elementary topology that Φ is a free group on two generators, say A and B. It follows that

$\Phi(N)$ = unique normal subgroup of Φ such that $\Phi/\Phi(N) \approx \mathbf{Z}(N)^2$, and an isomorphism is given by

$$\sigma \mapsto (\alpha(\sigma), \beta(\sigma)) \bmod N.$$

It follows also that

$$\Phi(N) \text{ is generated by } A^N, B^N \text{ and } \Phi^c.$$

Thus $\Phi(N)$ is the smallest normal subgroup containing $ABA^{-1}B^{-1}$, A_N, B_N; and therefore that $\Phi(N)$ is also generated by all elements

$$\gamma ABA^{-1}B^{-1}\gamma^{-1}, \text{ with } \gamma \in \Phi; \quad A^N \text{ and } B^N.$$

But the elements $\gamma ABA^{-1}B^{-1}\gamma^{-1}$ are precisely those of the form

$$\sigma\rho ABA^{-1}B^{-1}\rho^{-1}\sigma^{-1},$$

with $\sigma \in \Phi(N)$ and ρ ranging over a set of right coset representatives of $\Phi(N)$ in Φ. Hence we have proved:

Lemma 3.1. *A set of generators for* $\Phi(N)$ *mod* $\Phi(N)^c$ *is given by*

$$A^N, \quad B^N, \quad A^j B^k ABA^{-1}B^{-1}(A^j B^k)^{-1} \qquad \text{with} \qquad 0 \leqq j, k \leqq N - 1.$$

For the applications, we have to determine two generators A, B more precisely. We select a real point z_0 with $0 < z_0 < 1$, and we let $u \in U$ lie above z_0, i.e., $z_0 = \lambda(u)$. We pick a closed path starting at z_0, in $\mathbf{C} - [1, \infty)$, having winding number 1 around 0 and lift it to U, starting at the point u.

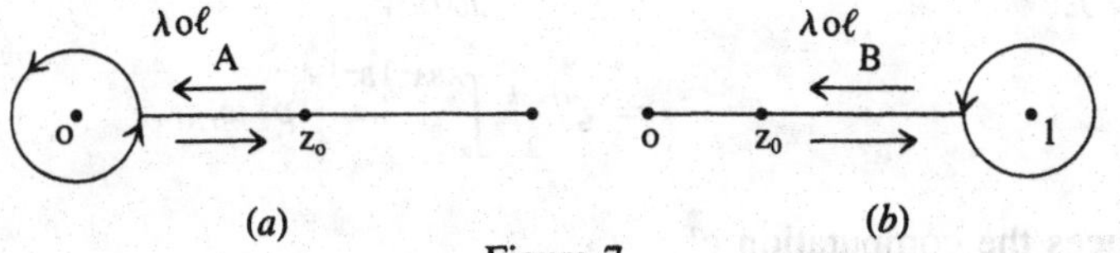

Figure 7

Then the lifted path ends at a point Au for some uniquely determined $A \in \Phi$. We denote this lifted path by l_A, so that $\lambda \circ l_A$ is homotopic to the path in $\mathbf{C} - \{0, 1\}$ shown in Fig. 7(a). Similarly, we construct l_B so that $\lambda \circ l_B$ is the path shown in Fig. 7(b). Then B is uniquely determined such that $\lambda \circ l_B$ is contained in

$$\mathbf{C} - (-\infty, 1],$$

and has winding number 1 around 1. We obtain:

$$L(\lambda, A) = 1, \qquad L(1 - \lambda, A) = 0$$
$$L(\lambda, B) = 0, \qquad L(1 - \lambda, B) = 1.$$

We know from Chapter II that the dfk's are of the form

$$p^*\omega_{r,s} = \widetilde{x}^{\,r-1} \widetilde{y}^{\,s-1} \frac{dx}{y^{N-1}} \quad \text{with} \quad 1 \leqq r, s \quad \text{and} \quad r + s \leqq N - 1.$$

Hence

$$\sigma^* p^* \omega_{r,s} = \zeta^{\alpha(\sigma)r + \beta(\sigma)s} p^* \omega_{r,s}.$$

We recall the definition of the beta function:

$$B(a, b) = \int_0^1 t^{a-1} (1 - t)^{b-1} dt.$$

Theorem 3.2 (Rohrlich). *The period lattice of $F(N)$ relative to the basis $\{\omega_{r,s}\}$ is spanned by the vectors*

$$(\ldots, \zeta^{rj + sk} (1 - \zeta^r)(1 - \zeta^s) \frac{1}{N} B\left(\frac{r}{N}, \frac{s}{N}\right), \ldots)$$

for $0 \leqq j, k \leqq N - 1$.

Proof. We follow that the general procedure of §2 with the representatives of Φ/Φ^c that we found above. We get:

$$\int_u^{A^j B^k ABA^{-1}(A^j B^k)^{-1}u} p^*\omega_{r,s} = \zeta^{rj+sk} \int_{(A^j B^k)^{-1}}^{ABA^{-1}B^{-1}(A^j B^k)^{-1}u} p^*\omega_{r,s}$$

$$= \zeta^{rj+sk} \int_u^{ABA^{-1}B^{-1}u} p^*\omega_{r,s}.$$

This reduces the computation of

$$\int_u^{\sigma u} p^*\omega_{r,s}$$

for representatives of $\Phi(N)/\Phi(N)^c$ to the cases

$$\sigma = ABA^{-1}B^{-1}, \qquad \sigma = A^N, \qquad \sigma = B^N.$$

First let $\sigma = ABA^{-1}B^{-1}$. Let $x^N = z$. Then

$$\int_u^{\sigma u} p^*\omega_{r,s} = \int_{\lambda \circ l_\sigma} (z^{1/N})^{r-1}[(1-z)^{1/N}]^{s-N} d(z^{1/N})$$

where

$$\lambda \circ l_\sigma = (\lambda \circ l_A)(\lambda \circ l_B)(\lambda \circ l_A)^{-1}(\lambda \circ l_B)^{-1},$$

and the differential form under the integral sign is the analytic continuation of the real principal value on the real segment between 0 and 1. Then $\lambda \circ l_\sigma$ is homotopic to the sum of four pieces as shown.

Figure 8

The circles have radius ϵ, tending to 0, so in the limit the final integral can be taken to be that over the segment, properly oriented, with the appropriate analytic continuations.

After the first turn around 0, the differential form changes by a factor of $\zeta^{r-1}\zeta = \zeta^r$. After the turn around 1, it changes by a factor of ζ^s, and thus becomes $\zeta^{r+s}\omega_{r,s}$. After the final turn around 0, it becomes $\zeta^s \omega_{r,s}$. Therefore:

$$\int_{\lambda \circ l_\sigma} \omega_{r,s}(z) = (-1 + \zeta^r - \zeta^{r+s} + \zeta^s) \int_0^1 t^{(r-1)/N}(1-t)^{(s-N)/N} d(t^{1/N}).$$

Making a change of variables in the integral on the right immediately gives the answer

$$- (1 - \zeta^r)(1 - \zeta^s) \frac{1}{N} B \left(\frac{r}{N}, \frac{s}{N} \right),$$

as desired.

If $\sigma = A^N$ or $\sigma = B^N$ then a similar argument shows that

$$\int_u^{\sigma u} p^* \omega_{r,s} = \int_{\lambda \circ l_\sigma} (z^{1/N})^{r-1} [(1 - z)^{1/N}]^{s-N} d(z^{1/N})$$

$$= 0,$$

because the value is equal to the Beta integral times

$$1 + \zeta + \cdots + \zeta^{N-1} = 0.$$

This proves the theorem.

§4. Periods on the Related Curves

In Chapter II, §3 we had defined the curves $F(r, s)$ by letting

$$u = x^N \quad \text{and} \quad v = x^r y^s.$$

Then $F(r, s)$ is the non-singular curve whose function field is

$$\mathbf{C}(u, v).$$

Let A, B be the generators of Φ which we selected previously, having winding numbers 0 and 1 respectively around 0 and 1. We let

$$\Phi(r, s) = \text{Aut}(U/F(r, s)'),$$

where $F(r, s)'$ is the affine curve obtained by deleting the points at infinity from $F(r, s)$. We have a Galois diagram:

$$
\begin{array}{ccc}
U & & \\
\downarrow & & \\
F(N)' & k(x, y) & \\
\downarrow & \mid & \left. \begin{array}{c} \\ \\ \\ \end{array} \right\} \Phi(r, s)/\Phi(N) \\
F(r, s)' & k(x^N, x^r y^s) & \\
\downarrow & \mid & \\
F(1)' & k(x^N, y^N) &
\end{array}
$$

Then $\Phi(r, s)$ is generated by $\Phi(N)$, A^M, B^M and $A^s B^{-r}$ where

$$M = N/D \quad \text{and} \quad D = \text{g.c.d.} \ (r, s, N).$$

Theorem 4.1 (Rohrlich). *The period lattice of $F(r, s)$ is generated by the vectors*

$$(\ldots, \zeta^{rmj+smk}(1 - \zeta^{rm})(1 - \zeta^{sm})\frac{1}{N}B\left(\frac{\langle rm \rangle}{N}, \frac{\langle sm \rangle}{N}\right), \ldots)$$

where $1 \leq j, k < M - 1$, components of the above vector are indexed by positive integers m which are (r, s)-admissible (Chapter II, §3), together with the element

$$(\ldots, (1 - \zeta^{srm})\frac{1}{N}B\left(\frac{\langle rm \rangle}{N}, \frac{\langle sm \rangle}{N}\right), \ldots).$$

Proof. The integral of $\omega_{\langle rm \rangle, \langle sm \rangle}$ over $l_A^s l_B^{-r}$ is the only new one, not already considered in determining the period lattice on the Fermat curve itself. The integral over l_A^s is obtained by integrating over the path shown, letting ϵ tend to 0.

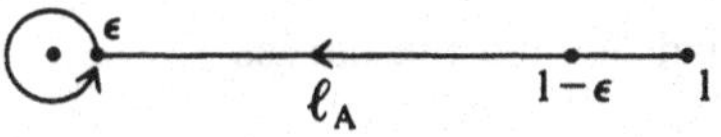

Figure 9

The analytic continuation of $\omega_{\langle rm \rangle, \langle sm \rangle}$ over l_A changes by a factor of ζ^{rm}, and this is repeated s times. Hence

$$\int_{l_A^s} \omega_{\langle rm \rangle, \langle sm \rangle} = (-1 + \zeta^{srm}) \int_0^1 \text{principal value,}$$

thus giving the extra element generating the period lattice.

CHAPTER VI

Linear Theory of Theta Functions

§1. Associated Linear Forms

Let V be a complex vector space of dimension n, real dimension $2n$. Let D be a lattice in V, that is, a discrete subgroup of real dimension $2n$, so that the factor group V/D is a complex torus. We define a **theta function** on V, with respect to D (or on V/D), to be a quotient of entire functions (called a meromorphic function for this chapter), not identically zero, and satisfying the relation

$$F(x + u) = F(x)e^{2\pi i[L(x,u)+J(u)]}, \qquad \text{all } x \in V, u \in D$$

where L is **C**-linear in x, and no specifications are made on its dependence on u, or on the dependence of the function J on u. However, we note that we can change J by a **Z**-valued function on D without changing the above equation. Also, we shall see below that any such L and J must satisfy additional conditions which can be deduced from this equation. We note that the theta functions form a multiplicative group.

Example. Let q be a quadratic form on V, such that

$$q(x) = B(x, x)$$

where B is symmetric and **C**-bilinear. Let c be a complex number, and let λ be a **C**-linear form on V. The function

$$e^{2\pi i[q(x)+\lambda(x)+c]}$$

is obviously a theta function, which will be called **trivial**. It is clear that the trivial theta functions form a multiplicative group. Furthermore, the trivial theta function above is a product of a constant, and the theta functions formed separately with $q(x)$ and $\lambda(x)$.

Note that $q(x + y) - q(x) - q(y) = 2B(x, y)$. Hence if

$$F(x) = e^{2\pi i q(x)},$$

then the associated functions L and J can be selected to be

$$L_F(x, u) = 2B(x, u) \quad \text{and} \quad J_F(u) = q(u).$$

If

$$F(x) = e^{2\pi i \lambda(x)},$$

then the associated functions L and J can be selected to be

$$L_F = 0 \quad \text{and} \quad J_F(u) = \lambda(u).$$

It will be useful later to multiply any theta function by a trivial one for normalization purposes.

Remark. Let F be an entire theta function which has no zero. Then we can write $F(z) = e^{2\pi i g(z)}$, where g is an entire function. From the definition of a theta function, we find that

$$g(z + u) - g(z) = L(z, u) + J(u).$$

All second order partial derivatives of this expression with respect to complex coordinates vanish, and hence the second partials of g are periodic entire functions, whence constants. It follows that g is a polynomial of degree at most 2, in other words that the theta function F is of the form considered in the preceding example.

We shall now investigate the relations satisfied by L and J for an arbitrary theta function F. Computing $F(x + u + v)/F(x)$ in the obvious ways, we find the following relations:

$$L(x, u + v) + J(u + v) \equiv L(x, u) + L(x + u, v) + J(u) + J(v)$$

$$(\bmod \mathbf{Z})$$

Putting first $x = 0$, we then find:

(1) $$J(u + v) - J(u) - J(v) \equiv L(u, v) \qquad (\bmod \mathbf{Z})$$

(2) $$L(u, v) \equiv L(v, u) \qquad (\bmod \mathbf{Z})$$

(3) $$L(x, u + v) \equiv L(x, u) + L(x, v) \qquad (\bmod \mathbf{Z})$$

The second relation comes from the fact that the expression in (1) is symmetric in u and v. The difference between the two sides of (3) is an integer. Being linear in x, it must vanish, and hence we can replace the congruence by an equality in (3). We can then extend $L(x, u)$ to a function $L(x, y)$ on $V \times V$ which is **C**-linear in x and **R**-linear in y. (As usual, we do it first for rational multiples of elements of D, and then extend to the whole space by continuity.)

Theorem 1.1. *Let*

$$E(x, y) = L(x, y) - L(y, x).$$

*Then E is **R**-bilinear, alternating, and real valued on $V \times V$. Furthermore, E takes on integral values on $D \times D$.*

Proof. The last statement follows from (2). Since L is **R**-bilinear, it follows that E is real valued, so our theorem is proved.

Theorem 1.2. *Let*

$$S(x, y) = E(ix, y).$$

Then S is symmetric. Also the form

$$H(x, y) = E(ix, y) + iE(x, y)$$

is hermitian, so S is the real part of H.

Proof. We expand the value for $S(x, y) - S(y, x)$ in terms of L. We find at once that

$$S(x, y) - S(y, x) = i[E(x, y) - E(ix, iy)].$$

Since the left-hand side is real and the right-hand side is pure imaginary, they must both be 0. Hence S is symmetric, and also

$$E(x, y) = E(ix, iy).$$

One proves that

$$H(ix, y) = iH(x, y), \qquad H(x, iy) = -iH(x, y),$$
$$H(x, y) = \overline{H(y, x)}$$

by direct computation. Thus H is hermitian, as was to be shown.

Two theta functions are called **equivalent** if their quotient is a trivial theta function. We are interested in finding a unique theta function in an equivalence class (up to a constant factor), determined by additional normalizing conditions.

Suppose we have two solutions L, L_1 to the equations:

$$L(x, y) - L(y, x) = E(x, y)$$
$$L_1(x, y) - L_1(y, x) = E(x, y).$$

Then $L - L_1$ is symmetric and **C**-linear in x, so **C**-bilinear. If we change F by $e^{2\pi i q(x)}$ where q is a quadratic form, then the associated L changes by a symmetric **C**-bilinear form. On the other hand, we have

$$E(x, y) = \frac{1}{2i} [H(x, y) - H(y, x)].$$

Therefore

$$L_1(x, y) = \frac{1}{2i} H(x, y)$$

is a possible solution, differing from L by a **C**-bilinear symmetric form. Changing F by a unique factor $e^{2\pi i q(x)}$ (up to a constant) makes it so that the associated L satisfies

$$L(x, y) = \frac{1}{2i} H(x, y).$$

For a further normalization, we still can multiply the function by the exponential of a **C**-linear term. Define

$$K(u) = J(u) - \frac{1}{2} L(u, u).$$

We shall find a **C**-linear function λ such that $\operatorname{Im} \lambda \equiv \operatorname{Im} K$ mod **Z**.
From (1) we get

(4) $$K(u + v) \equiv K(u) + K(v) + \frac{1}{2} E(u, v) \qquad (\text{mod } \mathbf{Z}).$$

This shows that $\operatorname{Im} K$ is additive on D, with values in **C**/**Z**. Since D is free, we can lift $\operatorname{Im} K$ to a **C**-valued function which is additive on D, and then extend it to an **R**-linear function on V. Let $g = \operatorname{Im} K$, and let

$$\lambda(x) = g(ix) + ig(x).$$

Then

$$\lambda(ix) = g(-x) + ig(ix)$$
$$= -g(x) + ig(ix) = i\lambda(x),$$

so that λ is **C**-linear, and $K - \lambda$ is real valued.

We may now state our desired normalization.

We shall say that a theta function is **normalized** if it satisfies the conditions:

N 1.
$$L(x, y) = \frac{1}{2i} H(x, y).$$

N 2. *The function K on V is real valued.*

It is clear that the normalized theta functions form a subgroup of all theta functions. For a **normalized theta function**, the basic relation takes the form

$$F(x + u) = F(x) \exp\left(2\pi i\left[\frac{1}{2i} H(x, u) + \frac{1}{4i} H(u, u) + K(u)\right]\right).$$

If we define a map $\psi\colon D \to \mathbf{C}_1$ (complex numbers of absolute value 1) by

$$\psi(u) = e^{2\pi i K(u)},$$

then the basic property (4) for K can be written

$$\psi(u + v) = \psi(u)\psi(v)e^{2\pi i E(u,v)/2}.$$

One sometimes calls ψ the associated **quadratic character**.

The previous discussion proves:

Theorem 1.3. *In any equivalence class of theta functions, there exists a normalized theta function, unique up to a non-zero constant factor.*

Proof. We have shown existence. As to uniqueness, we have already noted the uniqueness of the quadratic factor. For the linear factor, if Im $K = 0$ then the **C**-linear function λ in the previous discussion is 0, which shows the uniqueness.

In fact, the arguments prove more than that, because they give us an isomorphism between spaces of theta functions as follows. Let Th(L, J) as before denote the space of theta functions of type (L, J). On the other hand, if H is the hermitian form associated with L, and if ψ is a function on D of absolute value 1, defined by the above equation, let

Th$_{\text{norm}}$ (H, ψ) = space of theta functions which are normalized,

$$\text{with } L = \frac{1}{2i} H \quad \text{and} \quad \psi = e^{2\pi i K}.$$

Theorem 1.4. *Given (L, J) there exists a unique (up to constant factor) trivial theta function g such that the map*

$$F \mapsto gF$$

gives an isomorphism

$$\text{Th}(L, J) \xrightarrow{\approx} \text{Th}_{\text{norm}} (H, \psi).$$

The next theorem gives additional information in the case of entire theta functions.

Theorem 1.5. *Suppose that F is a normalized entire theta function. Then:*

(i)   *There exists a number $C > 0$ such that*

$$|F(z)| \leq C e^{(\pi/2)H(z,z)} \quad \text{for all } z \in V.$$

(ii)   *The associated hermitian form H is positive (not necessarily definite).*

Proof. As to the first inequality, let

$$g(z) = F(z)e(-(\pi/2)H(z, z)),$$

where $e(w) = e^w$. Then for $u \in D$,

$$g(z + u) = g(z)e^{i\pi[E(z,u)+K(u)]}.$$

The exponent in brackets is real. Hence $|g(z + u)| = |g(z)|$. This means that the function $|g|$ is periodic, and continuous, so that it is bounded. Our inequality follows at once.

Now suppose that $H(z_0, z_0) < 0$ for some complex number z_0. The function $z \mapsto F(zz_0)$ is entire, and from our inequality we see at once that it tends to 0 at infinity. Hence it is constant, and this constant must be 0. Since $H(z, z) < 0$ for all z sufficiently close to z_0, we see that F is equal to 0 in a neighborhood of z_0, whence F is identically 0, which is impossible. This proves our theorem.

§2. Degenerate Theta Functions

We shall see that the theory of theta functions factors through the subspace of V on which the hermitian form is trivial. Let V_H be the subset of V consisting of all z such that $H(z, z) = 0$. Then V_H is a complex subspace, called the **kernel** of H, or **null space** of H. We note that if $H(z, z) = 0$ then $H(z, w) = 0$ for all $w \in V$. Indeed, for real t we have

$$H(w + tz, w + tz) = H(w, w) + 2t \operatorname{Re} H(z, w).$$

Letting t be large shows that $\operatorname{Re} H(z, w) = 0$. Similarly, the imaginary part is also 0, using it instead of t. This proves both that V_H is a vector space, and that the weaker condition $H(z, z) = 0$ proves the stronger one as stated.

Thus H induces a hermitian form on V/V_H, and the above property shows that on V/V_H, H is definite, that is, it has no null space. If H is positive, then the induced form on V/V_H is positive definite.

Theorem 2.1. *Let F be a normalized entire theta function on the complex space V, with respect to the lattice D, and let H be its associated hermitian form. Let V_H be the kernel of H. Then:*

(i) *The image of the lattice D in V/V_H is discrete.*

(ii) *The values of F depend only on the cosets of V_H.*

Proof. Let $z_0 \in V_H$ and let $x \in V$. By assumption, for any complex z, we have

$$H(x + zz_0, x + zz_0) = H(x, x).$$

The estimate in Theorem 1.5 shows that

$$|F(x + zz_0)| \leqq Ce\left(\frac{\pi}{2} H(x, x)\right).$$

Hence $F(x + zz_0)$ is an entire, bounded function of z, whence constant, and therefore equal to $F(x)$. This proves (ii). As for (i), let $u_1, \ldots, u_r$ be elements of D whose residue classes $\bar{u}_i \pmod{V_H}$ generate V/V_H over $\mathbf{R}$. For all $x \in V$ sufficiently close to 0, we have

$$|E(x, u_i)| < \tfrac{1}{2},$$

and hence if v is an element of D such that $\bar{v}$ is close to 0, we must have $E(v, u_i) = 0$ for all i, because $E(v, u_i)$ is an integer. Therefore $E(v, z) = 0$ for all $z \in V$ and hence $H(v, z) = 0$ for all $z \in V$. Hence $v \in V_H$, and $\bar{v} = 0$. This proves that the image of D in V/V_H is discrete.

From Theorem 2.1 we may view F as a theta function on V/V_H with respect to the lattice $D \bmod V_H$. Its associated hermitian form on V/V_H is that induced by H, and is **positive definite**. This is the case which will be further analyzed.

§3. Dimension of the Space of Theta Functions

Again we consider a complex vector space V of dimension n, with a lattice D. By a **Riemann form** E on V with respect to D, we mean a real-valued bilinear form

$$E: V \times V \to \mathbf{R},$$

satisfying the following conditions:

RF 1. *The form E is alternating.*

RF 2. *It takes integral values on $D \times D$.*

RF 3. *The form $(x, y) \mapsto E(ix, y)$ is symmetric positive.*

If this last symmetric form is also positive definite, then we say that E is a **non-degenerate** Riemann form.

If we select a basis for D over $\mathbf{Z}$, then it is also a basis for V over $\mathbf{R}$. The matrix representing a non-degenerating Riemann form E with respect to such a basis has integer coefficients, and its determinant is a perfect square. The square root of this determinant is called the **pfaffian** of E **with respect to** D, and is independent of the choice of such a basis (the independence is clear since the matrix of the form with respect to different bases changes by an integral matrix and its transpose, of determinant 1). Actually the proof that this determinant is a square will come out of the forthcoming lemma, where we select a suitably normalized basis.

As a matter of notation, if $u_1, \ldots, u_m$ are elements of V, we denote by $[u_1, \ldots, u_m]$ the $\mathbf{Z}$-module generated by these elements. In all cases to arise, they will also be linearly independent over $\mathbf{Z}$. They will also be in D, and hence independent over $\mathbf{R}$.

Lemma 1. *Let E be an alternating non-degenerate bilinear form on a free $\mathbf{Z}$-module D, having values in $\mathbf{Z}$. Then D is an E-orthogonal direct sum*

$$D = [e_1, v_1] \oplus \cdots \oplus [e_n, v_n]$$

of 2-dimensional submodules $[e_j, v_j]$, such that $E(e_j, v_j) = d_j$ is an integer > 0, and $d_1 \mid d_2 \mid \cdots \mid d_n$.

Proof. The lemma is proved by induction. Among all values $E(u, v)$ with $u, v \in D$ we select a least positive one, say d_1, and we take a pair e_1, v_1 such that $E(e_1, v_1) = d_1$. We let $[e_1, v_1] = D_1$ be the 2-dimensional **Z**-module generated by e_1, v_1. Let $D_1^\perp$ be the orthogonal complement of $[e_1, v_1]$ with respect to E. Then $D_1 \cap D_1^\perp = \{0\}$, because the only element of D_1 perpendicular to both e_1 and v_1 is 0. To see that $D = D_1 + D_1^\perp$ we use the standard Gram-Schmidt orthogonalization process. Given $u \in D$ we can obviously solve for numbers a, b such that

$$u - ae_1 - bv_1$$

is orthogonal to $[e_1, v_1]$ (with respect to E). For instance,

$$E(u - ae_1 - bv_1, e_1) = E(u, e_1) + bd_1.$$

Since **Z** is principal, and d_1 is the positive generator of the ideal of all values of E, it follows that d_1 divides $E(u, e_1)$. Hence we can solve for b to make the above expression equal to 0. Similarly, we solve for a, and thus prove our lemma.

We observe that the lemma (due to Frobenius) gives the analogue of the existence of an orthogonal basis for symmetric forms. An orthogonal decomposition and a basis $[e_1, v_1, \ldots, e_n, v_n]$ as in the lemma will be called a **Frobenius decomposition** and a **Frobenius basis for** D, **with respect to** E.

Let

$$L : V \times V \to \mathbf{C}$$

be an **R**-bilinear map, **C**-linear in its first variable, and such that its **associated alternating form**

$$E(x, y) = L(x, y) - L(y, x)$$

is a Riemann form. Let

$$J : D \to \mathbf{C}$$

be a function satisfying (1), that is

$$J(u + v) - J(u) - J(v) \equiv L(u, v) \qquad (\text{mod } \mathbf{Z}).$$

Then we shall call (L, J) a **type** for theta functions, with respect to (V, D). It is obvious that the set of theta functions of type (L, J) (together with 0) forms a complex vector space denoted $\mathbf{Th}(\mathbf{L}, \mathbf{J})$. We are striving for a theorem which will give us the dimension of that space.

If E is a Riemann form with respect to (V, D), then we can define a hermitian form

$$H(x, y) = E(ix, y) + iE(x, y),$$

and then obtain a bilinear form

$$L(x, y) = \frac{1}{2i} H(x, y).$$

To get a type (L, J), we still have the freedom of choosing the function J, but at least we see that L and H determine each other. Of course, J is determined only mod $\mathbf{Z}$.

A theta function will be called **non-degenerate** if its associated form H is positive definite (or in other words, its associated Riemann form E is non-degenerate).

Lemma 2. *Let* $[e_1, v_1] \oplus \cdots \oplus [e_n, v_n]$ *be a Frobenius decomposition of D with respect to a non-degenerate Riemann form. Then* $\{e_1, \ldots, e_n\}$ *is a* **C**-*basis for* V.

Proof. Let V' be the **R**-space generated by $e_1, \ldots, e_n$, and let $V'' = iV'$. Suppose that we have a relation

$$x + iy = 0, \qquad x, y \in V'.$$

Then iy is in V' (because equal to $-x$), hence is E-orthogonal to V'. But

$$E(iy, y) > 0, \quad \text{if} \quad y \neq 0.$$

Hence $y = 0$ by the non-degeneracy condition. Hence $x = 0$. This proves our lemma.

Observe that the lemma shows that if V' is the space generated over **R** by $e_1, \ldots, e_n$ then

$$V = V' + iV',$$

and that this is an **R**-direct sum.

Let $B: V \times V \to \mathbf{C}$ be symmetric and $\mathbf{C}$-bilinear, and let

$$\lambda: V \to \mathbf{C}$$

be $\mathbf{C}$-linear. It is clear that the space of entire theta functions of type (L, J) is isomorphic to the space of entire theta functions of type $(L + B, J + \lambda)$. Indeed, if G is a trivial theta function of type (B, λ) (formed by exponentiating a quadratic form and a linear form), then the association

$$F \mapsto GF$$

gives an isomorphism between the two spaces, its inverse being given by multiplication with G^{-1}, which is of type $(-B, -\lambda)$. Hence to determine the dimension of the space

$$\mathbf{Th}(\mathbf{L}, \mathbf{J})$$

of entire theta functions of type (L, J), we shall be able to adjust our L, J to be most convenient. The dimension is now given by the following theorem.

Theorem 3.1 (Frobenius). *Let (L, J) be a non-degenerate type with respect to (V, D). Then the entire theta functions on V with respect to D having this type, together with 0, form a complex vector space of dimension equal to the pfaffian of E with respect to D.*

Proof. We first observe that L is symmetric on $[e_1, \ldots, e_n]$ (notation as above), and hence on the space V' generated by $e_1, \ldots, e_n$ over $\mathbf{R}$. Consequently, there is a symmetric $\mathbf{C}$-bilinear form B on V such that $L - B$ is 0 on $[e_1, \ldots, e_n]$, whence 0 on V'. Similarly, there is a $\mathbf{C}$-linear form λ on V such that $J - \lambda$ is 0 on $e_1, \ldots, e_n$. By the above remarks, this reduces our proof to the case when

(*) $L(z, e_j) = 0$ for all $z \in V$, and $J(e_j) = 0$ for all $j = 1, \ldots, n$.

In this case the conditions defining a theta function are particularly simple, and our theorem can be formulated as follows:

Theorem 3.2. *Let (L, J) be a non-degenerate type with respect to (V, D). Assume that*

$$D = [e_1, v_1] \oplus \cdots \oplus [e_n, v_n]$$

is a Frobenius decomposition, and that (*) holds. Let $c_j = J(v_j)$, and for $z \in V$ let $\{z_1, \ldots, z_n\}$ be the coordinates of z with respect to the $\mathbf{C}$-basis $\{e_1, \ldots, e_n\}$. Then the space of entire theta functions of type (L, J) is *precisely the space of entire functions satisfying the conditions*

$$F(z + e_j) = F(z)$$
$$F(z + v_j) = F(z)\mathbf{e}(z_j d_j + c_j),$$

where $\mathbf{e}(w) = e^{2\pi i w}$, *and this space has dimension* $d_1 \cdots d_n$, *where* $d_j = E(e_j, v_j)$.

Proof. It is clear that the two conditions are precisely those defining theta functions with respect to our lattice, and (L, J). We must now construct theta functions. In view of the conditions with respect to the e_j, we have periodicity, and hence any such theta function has a Fourier expansion

$$F(z) = \sum a(r)e^{2\pi i r \cdot z},$$

where the sum is taken over all $r \in \mathbf{Z}^n$. We shall also write $\langle r, z \rangle$ instead of the dot product $r \cdot z$. The second system of conditions now imposes relations among the coefficients $a(r)$, namely

$$a(r - d_j e_j) = a(r)\mathbf{e}(r \cdot v_j - c_j),$$

where again $\mathbf{e}(w) = e^{2\pi i w}$. The values $a(r)$ can be fixed arbitrarily for

$$0 \leq r_j < d_j \qquad (j = 1, \ldots, n)$$

and can be determined formally uniquely by the above relations for other values of r. This shows that the formal Fourier series solutions form a vector space of the required dimension $d_1, \ldots, d_n$.

There remains to prove convergence. Let us write

$$a(r) = e^{2\pi i g(r)} a(\bar{r})$$

where $\bar{r}$ is a representative $\text{mod}[d_1 e_1, \ldots, d_n e_n]$ satisfying the above inequalities. Then it suffices to solve the functional equation

$$g(r - d_j e_j) - g(r) + c_j = \langle r, v_j \rangle.$$

Using the fact that

$$L(e_i, v_j) = E(e_i, v_j)$$

because we have adjusted L so that $L(V, e_i) = 0$ for all i, it follows that if we write

$$v_j = \sum_k v_{jk} e_k,$$

then

$$L(v_j, v_k) = v_{jk} d_k = L(v_k, v_j),$$

this last equality being true because $E(v_j, v_k) = 0$. Hence

$$\langle r, v_j \rangle = \sum_k r_k d_k^{-1} L(v_j, v_k).$$

The above functional equation is essentially the polarization of a quadratic form, except for certain constants. Therefore the solution $g(r)$ can be expressed as

$$g(r) = q(r) + \text{linear term in } r + \text{constant},$$

where $q(r)$ is a quadratic form, which we can easily guess, namely

$$q(r) = -\frac{1}{2} L\left(\sum_k r_k d_k^{-1} v_k, \sum_k r_k d_k^{-1} v_k \right).$$

$$= -\frac{1}{2} L(w, w) \qquad \text{where } w = \sum_k r_k d_k^{-1} v_k.$$

Therefore, to obtain convergence, it will suffice to prove:

Lemma 3.3. *If L is such that $L(V, e_i) = 0$ for all i, that is, L satisfies* (*), *then the imaginary part of L is negative definite on the **R**-space generated by $v_1, \ldots, v_n$.*

Proof. Let $z = x + iy$, where x, y lie in the space V' above, that is are real linear combinations of $e_1, \ldots, e_n$. Then for $y \neq 0$ we get

$$0 < E(iy, y) = E(x + iy, y) = E(z, y) = L(z, y) - L(y, z) = -L(y, z)$$

by our assumption on L. However,

$$L(z, z) = L(x, z) + iL(y, z),$$

and $L(x, z) = E(x, z)$ is real, again by that same assumption. Hence if $y \neq 0$ we see that $\operatorname{Im} L(z, z) < 0$.

If $z = w$ as in the theorem, then z lies in the **R**-space generated by $v_1, \ldots, v_n$, and we cannot have $y = 0$, otherwise z also lies in V' which is impossible. This concludes the proof of Theorem 3.2.

Remark 1. In terms of matrices, we can express the negative definiteness as follows. Let T be the matrix such that

$$\frac{1}{d_j} v_j = T e_j \quad \text{for} \quad j = 1, \ldots, n,$$

viewing vectors as column vectors. In other words,

$$T = (v_1/d_1, \ldots, v_n/d_n).$$

Then T is symmetric, and its imaginary part is negative definite.

The proof simply consists in observing that if $T = (\tau_{jk})$, then

$$L(v_j, v_k) = d_j d_k \tau_{jk},$$

as follows at once from the linearity of L in the first variable. Furthermore, we have

$$\langle Tr, r \rangle = L(w, w),$$

where r is any real linear combination of $e_1, \ldots, e_n$ (that is, an element of V'), and w is defined above,

$$w = \sum r_k d_k^{-1} v_k.$$

Suppose that we started with a type (L, J) for which the alternating form E is degenerate. We are still interested in the dimension of the space of theta functions of this type. We may assume that it is normalized. Then the elements of $\mathrm{Th}(L, J)$ induce theta functions on the factor torus $\overline{V}/\overline{D}$, where $\overline{V} = V/V_H$, and V_H is the null space of the associated hermitian form. Furthermore

$$\dim \mathrm{Th}(L, J) = \dim \mathrm{Th}(\overline{L}, \overline{J}),$$

where $\overline{L}, \overline{J}$ is the induced type on $\overline{V}/\overline{D}$. Now $\overline{E}$ is non-degenerate, and the pfaffian of $\overline{E}$ will be called the **reduced pfaffian** of E. We then obtain:

Theorem 3.4. *The dimension of* $\mathrm{Th}(L, J)$ *is equal to the reduced pfaffian of the associated alternating form* E.

Proof. This follows from the theorem, which gives us the dimension in the non-degenerate case.

Remark 2. Let (L, J) be a non-degenerate type with respect to (V, D). It may happen that there is a bigger lattice D' for which (L, J) is also a nondegenerate type with respect to (V, D'). However, such D' are limited. Indeed, let $u \in D'$, and write

$$u = \sum_{j=1}^{n} a_j e_j + \sum_{j=1}^{n} b_j v_j$$

in terms of a Frobenius basis of V. Then

$$E(e_j, u) = b_j d_j \quad \text{and} \quad E(v_j, u) = -a_j d_j.$$

These values have to be integers. Hence a_j, b_j can take on only a finite number of values mod $\mathbf{Z}$, and there are only a finite number of possible lattices D'.

For each such D', the factor group D'/D is finite. By the functional equation (1) of §1, there are only a finite number of extensions of J to maps J' on such D'.

Theorem 3.5. *Let (L, J) be a non-degenerate type with respect to (V, D). Then all theta functions of this type except possibly those lying in a finite union of subspaces of dimension $< \mathrm{pf}(E)$ are not theta functions with respect to a lattice strictly larger than D.*

Proof. If F is a theta function of type (L, J') with respect to (V, D'), then the pfaffian of E with respect to D' is equal to

$$\frac{d}{(D' : D)} < d,$$

where $d = d_1 \cdots d_n$ is its pfaffian with respect to D. Hence the space of theta functions of type (L, J') with respect to (V, D') has lower dimension, and there is only a finite number of such spaces, for lattices D' properly containing D.

§4. Abelian Functions and Riemann-Roch Theorem on the Torus

By an **abelian function** on V with respect to D (or on the torus $T = V/D$) we shall mean a quotient of theta functions of the same type, or 0. It is then clear that the abelian functions form a field, called the **function field** of T, and denoted by $\mathbf{C}(T)$. Note that an abelian function is genuinely periodic with respect to D, that is

$$f(z + u) = f(z) \qquad z \in V, u \in D.$$

It follows immediately from the definitions that an abelian function $\neq 0$ is a quotient of entire theta functions of the same type.

If θ_0 is an entire theta function of type (L, J), we denote by $\mathcal{L}(\theta_0)$ the space of all entire theta functions of the same type. If θ lies in this space, then θ/θ_0

is an abelian function. When (L, J) is non-degenerate, then the dimension of $\mathcal{L}(\theta_0)$ is given by Theorem 3.1 of §3. We call $\mathcal{L}(\theta_0)$ the **linear system** of θ_0.

Two theta functions will be called **linearly equivalent** if their quotient is an abelian function, and this relation is denoted by

$$\theta \sim \theta'.$$

Recall Theorem 1.3 which states that in any equivalence class of theta functions (the equivalence being that of differing by a trivial theta function) there is precisely one normalized theta function.

Remark. Two normalized theta functions are linearly equivalent if and only if they have the same type.

The proof is immediate, because the quotient of normalized theta functions is normalized, and is an abelian function if and only if it is of type $(0, 0)$.

Let V_0 be the intersection of all the null spaces of all Riemann forms on (V, D). We call V_0 the **degenerate subspace of V with respect to D**. Then by taking a finite sum of Riemann forms on (V, D) we find that V_0 is itself the null space of a Riemann form on (V, D).

Since an abelian function is a normalized theta function, we can express it uniquely (up to a constant factor) as a quotient of normalized entire theta functions, which are in fact defined on V/V_0 by the result of §2. Consequently the function field of V/D is "the same" as the function field of $\overline{V}/\overline{D}$, where $\overline{V} = V/V_0$ and $\overline{D}$ is the image of D in $\overline{V}$. Furthermore, $(\overline{V}, \overline{D})$ is non-degenerate, that is, admits a non-degenerate Riemann form. Thus the study of the function field is reduced easily to that of the non-degenerate case.

At this point it is useful to introduce the terminology of divisors. In algebraic or complex analytic geometry, a divisor is a formal linear combination of irreducible subvarieties of codimension 1. Locally in the neighborhood of each point, such a variety is defined by a single equation $\varphi = 0$, where φ is some analytic function. It is clear that locally, if the subvariety is defined by another equation $\psi = 0$, then $\varphi\psi^{-1}$ is holomorphic invertible in the neighborhood of the point. It turns out that on $\mathbf{C}^n$, a positive divisor can be defined by a global entire theta function, and two such functions differ by an entire invertible function which is therefore a trivial theta function. However, for the applications we have in mind, the theorem giving the representation of a divisor by a theta function is utterly irrelevant: all we need are global considerations. Hence we shall prove the representation theorem in the last chapter, and here we make the relevant global definition, which still allows us to use the standard language of divisors. Namely, for our purposes, we define a **divisor** to be an equivalence class of theta functions, or 0. We shall write divisors additively. A divisor is said to be **linearly equivalent to zero** if it is the divisor of an abelian function, that is if its equivalence class contains an abelian function. Two divisors are said to be **linearly equivalent** if their difference is linearly equivalent to zero. We let Cl denote **linear equivalence**.

We may also introduce the notion of **ordering** between divisors. We say that $X \geqq 0$ if it is the equivalence class of an entire theta function. Similarly, $X \geqq Y$ if $X - Y \geqq 0$. In terms of theta functions, this means that θ_X/θ_Y is entire, if θ_X and θ_Y are theta functions in the equivalence classes of X and Y respectively. We let (θ) denote the divisor of θ.

Suppose that X_0 is a positive divisor. We denote by $\mathscr{L}(X_0)$ the space of abelian functions f such that

$$(f) \geqq -X_0.$$

If θ_0 is a theta function having divisor X_0, then it is clear that $\mathscr{L}(X_0)$ is isomorphic to the space $\mathscr{L}(\theta_0)$ defined previously under the isomorphism

$$\theta \mapsto \theta/\theta_0, \text{ for } \theta \in \mathscr{L}(\theta_0).$$

We say that X_0 is **non-degenerate** if a theta function defining X_0 is non-degenerate. If H, E are the hermitian and Riemann forms associated with θ_0, then we also say that they are **associated with the corresponding divisor**, and write $E = E_{X_0}$.

Observe that if E_0 is a non-degenerate Riemann form, and E is any Riemann form, then $E + E_0$ is non-degenerate, because the sum of a positive definite and a positive form is positive definite. In terms of divisors, if X_0 is positive non-degenerate, and X is positive, then $X + X_0$ is positive non-degenerate.

As in the case of curves, we let

$$l(X) = \dim \mathscr{L}(X).$$

Theorem 4.1 (Riemann-Roch). *Let $X_0, \ldots, X_m$ be positive divisors such that X_0 is non-degenerate. There exists a polynomial P in $m + 1$ variables such that*

$$l(r_0 X_0 + \cdots + r_m X_m) = P(r_0, \ldots, r_m)$$

for any integers $r_0, \ldots, r_m \geqq 0$ and $r_0 > 0$. For any non-degenerate positive divisor X with Riemann form E, we have

$$l(rX) = r^n \text{ pf}(E).$$

Proof. This is obvious from Theorem 3.1 because the desired dimension is given as the pfaffian of $r_0 E_0 + \cdots + r_m E_m$.

Corollary 1. *The function field of V/D has transcendence degree at most $n = \dim_C V$.*

Proof. Suppose that there exist abelian functions $f_1, \ldots, f_m$ which are algebraically independent over C. We can write them all over a common denominator θ_0, which we may assume non-degenerate (after factoring out the null space of V/D if necessary). Write

$$f_j = \theta_j / \theta_0.$$

Then the monomials

$$\theta_1^{r_1} \cdots \theta_m^{r_m}, \qquad r_1 + \cdots + r_m = r,$$

lie in $\mathscr{L}(\theta_0^r)$, whose dimension is given by Theorem 4.1 and is equal to $r^n \operatorname{pf}(E_0)$. However, the above monomials are linearly independent, and there are

$$\binom{m + r}{m}$$

of them, which grows like r^m. Hence $m \leq n$, as desired.

In §6, it will be proved that the transcendence degree of the function field is precisely n if V/D has a non-degenerate Riemann form. Under this assumption, we shall now use an argument similar to the above to prove that the field must then be finitely generated, that is, must be a finite extension of a purely transcendental extension.

Corollary 2. *If there exists n algebraically independent abelian functions on V/D, say $f_1, \ldots, f_n$, then the function field of V/D is a finite extension of $C(f_1, \ldots, f_n)$.*

Proof. Write again $f_j = \theta_j / \theta_0$. Then θ_0 must be non-degenerate, for otherwise, since θ_j and θ_0 have the same type, and can be assumed to be normalized, we could induce our functions on the factor space $\overline{V} = V/V_H$, where V_H is the null space of the hermitian form associated with θ_0, thus contradicting Corollary 1. Let g be any abelian function. Consider all monomials

$$f_1^{r_1} \cdots f_n^{r_n} g^s$$

with $r_1 + \cdots + r_n = r$. Each such monomial lies in $\mathscr{L}(rX_0 + sY)$ where $X_0 = (\theta_0)$ and $g \in \mathscr{L}(Y)$ with $Y > 0$. The number of such monomials is equal to

$$s\begin{pmatrix} r + n \\ n \end{pmatrix},$$

which is asymptotic to $sr^n/n!$ as $r \to \infty$. Since the dimension of $\mathscr{L}(rX_0 + sY)$ is equal to

$$\mathrm{pf}\,(rE_0 + sE_Y) = r^n \,\mathrm{pf}\,(E_0) + \text{lower terms},$$

it follows that if $s > n!\,\mathrm{pf}(E_0)$, then there exists a relation of linear dependence among these monomials. Such a relation gives an algebraic equation
for g, with coefficients in $\mathbf{C}\,[f_1, \ldots, f_n]$, of degree $\leq n!\,\mathrm{pf}(E_0)$. If we
select an abelian function of maximal degree over the field $\mathbf{C}\,(f_1, \ldots, f_n)$,
then by the primitive element theorem of elementary field theory, this abelian
function must generate the whole field of abelian functions, thereby proving
our theorem.

§5. Translations of Theta Functions

We now consider the type associated with translations of theta functions. Let
$a \in V$. If θ is a theta function, we let θ_a be the function such that

$$\theta_a(x) = \theta(x - a),$$

and call θ_a the **translation** of θ by a. It is obviously a theta function. The
definition is made such that if X is the divisor of θ, then X_a is the divisor of θ_a.

Lemma 5.1. (i) *If θ is of type (L, J) then θ_a is of type $(L, J - L_a)$, where*
$L_a(u) = L(a, u)$.

(ii) *If θ is normalized, the type of the normalized theta function in the same
equivalence class as θ_a is $(L, J - E_a)$ where $E_a(u) = E(a, u)$.*

Proof. We have

$$\frac{\theta(x - a + u)}{\theta(x - a)} = \exp\,\{2\pi i\,[L(x - a, u) + J(u)]\}$$

from which our first assertion is obvious. *Note that the bilinear form L is the
same for any translation of θ, whence the associated hermitian form H is the
same.*

For the second assertion, if we multiply θ_a by the trivial theta function

$$\exp\left\{2\pi i\left[\frac{1}{2i}\,H(x, a)\right]\right\},$$

then we get rid of the imaginary part of $-L(a, u)$ in the characterizing equation, remaining only with the real part in the exponent, given by

$$K(u) - E(a, u).$$

Theorem 5.2. *Let* θ, θ' *be entire normalized theta functions, with associated hermitian forms* H, H', *respectively. Then* $H = H'$ *if and only if there exists* $a \in V$ *such that the normalization of* $\theta_a \sim \theta'$.

Proof. Suppose such a exists. Since the type of θ_a is $(L, J - L_a)$, it follows that $L = L'$, whence $H = H'$, using the uniqueness of a normalized theta function. Conversely, suppose that $H = H'$. Let V_H be the null space of H. Then both θ and θ' induce theta functions on $V/V_H = \overline{V}$, by §2, and we have the formulas

$$\frac{\theta(x + u)}{\theta(x)} = \mathbf{e}[L(x, u) + \tfrac{1}{2}L(u, u) + K(u)],$$

$$\frac{\theta'(x + u)}{\theta'(x)} = \mathbf{e}[L(x, u) + \tfrac{1}{2}L(u, u) + K'(u)]$$

with the same L, where $\mathbf{e}(w) = e^{2\pi i w}$. Note that $K' - K \pmod{\mathbf{Z}}$ is a homomorphism by (4), §1, and that

$$e^{2\pi i[K'(u)-K(u)]}$$

depend only on the class $\overline{u}$ of u mod V_H. Therefore there exists an element $\overline{a} \in \overline{V}$ such that the character $e^{2\pi i[K'(u)-K(u)]}$ is given by

$$\mathbf{e}[K'(u) - K(u)] = e^{-2\pi i E(a,u)}.$$

We now use the functional equation for θ and θ'. Let $f = \theta'/\theta_a$. Then the hypothesis that θ, θ' are normalized (i.e. that

$$L(x, y) = \frac{1}{2i} H(x, y))$$

implies at once that $f(x + u) = f(x)$ for all u in D, or in other words that f is abelian. This proves the desired property.

In terms of divisors, we can express Theorem 5.2 as follows.

Let X, Y *be two positive divisors. Then* $E_X = E_Y$ *if and only if there exists* $a \in V$ *such that* $Y = X_a$.

If θ is a theta function, we let $\mathrm{Cl}(\theta)$ denote its class modulo the group generated by the trivial theta functions and the abelian functions. The factor group of theta functions modulo trivial ones, and abelian ones, will be called the **Picard group** of (V, D), or of V/D.

Two theta functions are called **algebraically equivalent** if one is the translation of the other (modulo trivial and abelian theta functions). In light of Theorem 5.2 we see that this condition is equivalent with the property that the associated Riemann forms (or hermitian forms) are equal. Furthermore, the notion of algebraic equivalence applies also to divisors, via representative theta functions. Thus a divisor may be said to be **algebraically equivalent to** 0 if and only if it satisfies either one of the following two equivalent conditions:

$X = Y - Y_a$ *for some positive divisor Y and some $a \in V$.*

The associated Riemann form is equal to 0.

It is obvious that the map

$$\varphi_\theta : a \mapsto \mathrm{Cl}(\theta_a/\theta) \quad \text{or} \quad \mathrm{Cl}(X_a - X)$$

is a homomorphism of the torus V/D into the Picard group, and in fact into the subgroup of elements algebraically equivalent to zero, which is denoted by $\mathrm{Pic}_0(V/D)$. We shall be interested in its kernel.

Theorem 5.3. *Let θ be an entire non-degenerate theta function with Riemann form E. Then the kernel of φ_θ is finite in V/D, and is represented by those elements $a \in V$ such that $E(a, u) \in \mathbf{Z}$ for all $u \in D$. We have*

$$\text{order of Ker } \varphi_\theta = \mathrm{pf}(E)^2 = (d_1 \cdots d_n)^2.$$

where $d_1, \ldots, d_n$ are the elementary divisors of E.

Proof. We have seen in Lemma 5.1 that if (L, J) is the type of θ, which we may assume to be normalized, then the type of the normalized theta function in the equivalence class of θ_a is $(L, J - E_a)$. Suppose that a is in the kernel of φ_θ. Then θ_a/θ is equivalent to an abelian function. The type of a normalized theta function in its equivalence class is then $(0, -E_a)$ by the above (depending on Lemma 5.1), and it is also $(0, 0)$ since an abelian function is normalized. This implies that

$$E_a(u) = E(a, u) \in \mathbf{Z}$$

for all $u \in D$. But E is $\mathbf{Z}$-valued on $D \times D$. Hence it follows that if we express a as a linear combination of a basis $\{u_1, \ldots, u_{2n}\}$ for D with real coefficients, then in fact these coefficients are rational and have bounded denominators. This means precisely that the kernel of φ_θ is finite in V/D. Using a Frobenius basis, one sees in fact that the denominators are the elementary divisors $d_1, \ldots, d_n$.

If θ_a/θ is a trivial theta function, then a fortiori it is equivalent to the constant abelian function 1, and the above assertion applies to prove the last statement.

Theorem 5.4. *Let θ be an entire theta function, normalized, with associated hermitian form H. Let V_H be the null space of H. Then $V_H + D$ is of finite index in the kernel of φ_θ, in V.*

Proof. This results at once from the non-degenerate case. Indeed, we know from §2 that θ induces a theta function on $\overline{V} = V/V_H$, with respect to the image $\overline{D}$ of D in V/V_H. If $\theta'_a \sim \theta$, and θ'_a is the normalized theta function equivalent to θ_a, then θ'_a is of type $(L, J - E_a)$ by Lemma 5.1. Since θ'_a/θ is an abelian function, it follows that $E(a, D) \subset \mathbf{Z}$. But $E(a, D) = E(\overline{a}, \overline{D})$. Thus $\overline{a}$ is in the kernel of $\varphi_{\overline{\theta}}$, in the torus $\overline{V}/\overline{D}$. We can then apply Theorem 5.3 to conclude the proof.

§6. Projective Embedding

Let θ_0 be an entire theta function, and let $\{\theta_0, \theta_1, \ldots, \theta_m\}$ be a basis for $\mathcal{L}(\theta_0)$. Then we may view this basis as giving a map

$$F: x \mapsto (\theta_0(x), \ldots, \theta_m(x))$$

of V/D into projective space $\mathbf{P}^m$, defined at all those points x where not all θ_j vanish simultaneously, and called the **map induced by the linear system** of θ_0. We shall see that if the torus has a non-degenerate Riemann form, then there exists θ_0 such that the above map gives a complex analytic embedding of the torus into projective space.

Instead of a basis for $\mathcal{L}(\theta_0)$, we could just as well take a set of generators for the vector space $\mathcal{L}(\theta_0)$. We would obtain a map defined at precisely the same points. It is also useful to remark that the map is defined at x if and only if there exists some θ in $\mathcal{L}(\theta_0)$ such that $\theta(x) \neq 0$. If θ is written as a linear combination of basis elements, the condition that $\theta(x) = 0$ is equivalent to the condition that the image of the point lie in a certain hyperplane.

Let θ be an entire theta function. Let X be the set of its zeros, that is, the set of points x such that $\theta(x) = 0$. We may view X as a subset of V, or as a subset of V/D, because it is clear that X is invariant under translations by elements of D. We denote by X^- the set of all elements $-x$, with $x \in X$. It is clear that X^- is the set of zeros of the theta function θ^-, that is, the function such that $\theta^-(z) = \theta(-z)$. Also, for any point $a \in V$, the translation X_a consisting of all points $x + a$ with $x \in X$ is the set of zeros of θ_a (this is the reason for defining θ_a as we did, by the formula $\theta_a(z) = \theta(z - a)$). The union of a finite number of such zero sets $X_1 \cup \cdots \cup X_m$ is the set of zeros of the product $\theta_1 \cdots \theta_m$ of the corresponding theta functions. In particular, it is not the whole space.

Theorem 6.1 (Lefschetz). *Let θ be an entire non-degenerate theta function. Then the map of V/D into projective space induced by the linear system $\mathcal{L}(\theta^3)$ is everywhere defined, and is an analytic embedding of V/D into projective space.*

Proof. For any points $a, b \in V$, we observe that the function

$$\theta(x - a)\theta(x - b)\theta(x + a + b)$$

lies in $\mathcal{L}(\theta^3)$. To see that the map is well defined, it suffices to prove that given a point x, there exist a, b such that the above product is not equal to 0. But this is trivial: We first find a such that $x - a$ does not lie in the set of zeros of θ, and then find b such that $\theta(x - b)\theta(x + a + b) \neq 0$. In each case this amounts to finding a point not lying in a finite union of sets of zeros of theta functions.

Next we prove that the map is injective. In other words, given $x, y \in V$, if x and y have the same image in projective space, then x and y differ by a lattice point. That the image of x and y is the same means that there exists a complex number $\gamma \neq 0$ such that for all b, z, and all F of the same type as θ, we have

$$F(x - z)F(x - b)F(x + z + b) = \gamma F(y - z)F(y - b)F(y + z + b).$$

By Theorem 3.5 we can select F in $\mathcal{L}(\theta)$ such that F is not a theta function with respect to any lattice strictly larger than D. Let $D' = D + \mathbf{Z}v$, where $v = x - y$. We shall prove that v is of finite order modulo D, and that F is a theta function with respect to D'. It then follows that $v \in D$, as desired.

Given any point z_0, we can find b such that

$$F(x - b)F(x + z_0 + b)F(y - b)F(y + z_0 + b) \neq 0,$$

and hence such that this inequality holds in a neighborhood of z_0. This means that in the neighborhood of z_0 there is a holomorphic function g_0 having no zero, and such that

$$F(x - z) = F(y - z)g_0(z)$$

in the neighborhood of that point. It is then clear that such functions g_0 are analytic continuations of each other, and therefore that there exists an entire function g without zeros such that for all z we have

$$F(x - z) = F(y - z)g(z).$$

With $v = x - y$, this formula can be written in the form

$$F(z + v) = F(z)h(z),$$

for some entire function h without zeros. The theta relation for F now shows that h is in fact a trivial theta function, of the form

$$h(z) = Ce^{2\pi i\lambda(z)},$$

where λ is **C**-linear. Furthermore v is in the kernel of φ_F and is therefore of finite order with respect to D by Theorem 5.3. We can easily determine λ as follows. Note that h is of type $(0, L_v)$ if (L, J) is the type of F, and $\lambda(u) - L(v, u) \in \mathbf{Z}$ for all $u \in D$. But

$$\lambda(u) - L(v, u) = \lambda(u) - L(u, v) - E(v, u).$$

Hence $\lambda(z) - L(z, v)$ is real valued because λ, L are **R**-linear in z, and the elements of D generate V over **R**. But λ and L are also **C**-linear, and consequently we must have

$$\lambda(z) = L(z, v).$$

It follows that F is a theta function with respect to D', and therefore concludes the proof.

There remains to see that the differential of our projective mapping does not vanish at any point, whence our map is an embedding since V/D is compact. Select any point x, and a function G in $\mathscr{L}(\theta^3)$ such that $G(x) \neq 0$. Any $F \in \mathscr{L}(\theta^3)$ gives rise to a possible projective coordinate, such that the corresponding affine coordinate when we dehomogenize with respect to G is F/G. To see that the differential of our mapping does not vanish at x, it suffices to prove that given a vector $v \in V$, $v \neq 0$, there exists one such function F such that

$$d(F/G)(x)v \neq 0.$$

We may take a basis for V such that $v = (1, 0, \ldots, 0)$. We have

$$d(F/G)(x) = \frac{G(x)dF(x) - F(x)dG(x).}{G(x)^2}$$

Suppose that for every $F \in \mathscr{L}(\theta^3)$, we have $d(F/G)(x)v = 0$. Then

$$\frac{dF(x)}{F(x)}v = \frac{dG(x)}{G(x)}v = \alpha$$

for a fixed number α, and every F such that $F(x) \neq 0$. In our coordinate system, we have

$$dF(x)v = \frac{\partial F}{\partial z_1}(x), \qquad \frac{dF(x)}{F(x)} v = \frac{1}{F(x)} \frac{\partial F}{\partial z_1}(x).$$

Let

$$f(z) = \frac{1}{\theta(z)} \frac{\partial \theta}{\partial z_1},$$

wherever defined. We select for F the function

$$F(z) = \theta(z - a)\theta(z + a + b)\theta(z - b),$$

with a, b arbitrary. Then we have

$$f(x - a) + f(x - b) + f(x + a + b) = \alpha$$

for all a, b outside an exceptional set where the denominators on the left vanish. Now we consider the function of z given by

$$f(x - z) + f(x - b) + f(x + z + b),$$

which is constant, and differentiate with respect to each variable z_j. We get

$$\frac{\partial f}{\partial z_j}(x - z) = \frac{\partial f}{\partial z_j}(x + z + b).$$

From the right-hand side, we see that these partial derivatives are constant for some open set of z, whence it follows that in an open set of z where f is defined, we have

$$\frac{1}{\theta(z)} \frac{\partial \theta}{\partial z_1} = \alpha_1 z_1 + \cdots + \alpha_n z_n + \beta,$$

with constants $\alpha_1, \ldots, \alpha_n, \beta$. Let

$$q(z) = \tfrac{1}{2}\alpha_1 z_1^2 + \alpha_2 z_1 z_2 + \cdots + \alpha_n z_1 z_n + \beta z_1.$$

Then the first partial derivative of the function

$$\theta(z)e^{-q(z)} = \theta_1(z)$$

is equal to 0 in some open set, whence everywhere. This means that θ_1 (which is equivalent to θ) depends only on $n - 1$ variables. But θ_1 is an entire nondegenerate theta function, and we have a contradiction of Theorem 5.3 and Theorem 5.4, thereby proving our theorem.

Corollary. *There exist n algebraically independent abelian functions on a torus V/D having a non-degenerate Riemann form.*

Proof. For any point on the torus, we can find abelian functions $f_1, \ldots, f_n$ which are analytic local coordinates at the point, that is, map a neighborhood of the point on an open set in $\mathbf{C}^n$. These functions are obviously algebraically independent (even analytically independent).

The above corollary completes the proof of the hypothesis we needed in §4 to see that the function field was a finitely generated extension of a purely transcendental extension in n variables.

If D is a lattice in V and $A = V/D$, then define A to be an **abelian manifold** if there exists a projective embedding of A. The above theorem proves:

Theorem 6.2. *If (V, D) admits a non-degenerate Riemann form, then V/D is an abelian manifold.*

The converse can also be proved (see Chapter X, §3). We do not need it for the rest of this book. Consequently for our purposes, it is more convenient to take as **definition** that A is an **abelian manifold** if and only if there exists a non-degenerate Riemann form on (V, D).

Appendix: The 1-dimensional Case

It may be useful to the reader if we recall here briefly the situation in dimension 1. To deal with theta functions in this case, we need only use one basic theta function, having a simple zero at all the lattice points. Let

$$D = [\omega_1, \omega_2].$$

Because of the one-variable situation, we can write down the function, namely the Weierstrass sigma function,

$$\sigma(z) = z \prod_{\omega \in D'} \left(1 - \frac{z}{\omega} \right) \exp \left[\frac{z}{\omega} + \frac{1}{2} \left(\frac{z}{\omega} \right)^2 \right],$$

where D' is the lattice from which 0 has been deleted. Taking the logarithmic derivative yields what is known as the Weierstrass zeta function, namely

$$\zeta(z) = \frac{1}{z} + \sum_{\omega \in D'} \left[\frac{1}{z - \omega} + \frac{1}{\omega} + \frac{z}{\omega^2} \right].$$

Taking the derivative once more yields minus the Weierstrass $\wp$-function, and one sees easily that it is periodic. Hence for any period ω we get

$$\zeta(z + \omega) = \zeta(z) + \eta(\omega),$$

where η is **Z**-linear in ω. Integrating, and exponentiating, it follows therefore that

$$\sigma(z + \omega) = \psi(\omega) \exp\left[\eta(\omega)\left(z + \frac{\omega}{2} \right) \right] \sigma(z),$$

with a suitable function $\psi(\omega)$. Note here that

$$L(z, \omega) = \eta(\omega)z$$

is **C**-linear in z, and **Z**-linear in ω, as it should. Thus we see that σ is a theta function. It is easy to prove that

$$\psi(\omega) = \quad 1 \quad \text{if} \quad \omega/2 \quad \text{is a period,}$$
$$\psi(\omega) = -1 \quad \text{if} \quad \omega/2 \quad \text{is not a period.}$$

Finally, to get a theta function associated with a divisor, we just take translations of σ, and products, for example,

$$\prod_{i=1}^{r} \sigma(z - a_i)^{m_i},$$

with suitable multiplicities. Thus the theorem to be proved in Chapter IX becomes essentially trivial in dimension 1, because of the Weierstrass product expression for entire functions.

Consider the linear system of divisors $\geq -3(0)$, where 0 is the origin. In that system, we have the Weierstrass $\wp$-function

$$-\wp(z) = \zeta'/\zeta(z)$$

and its derivative $\wp'(z)$. The general theorem asserts that the coordinates

$$(1, \wp(z), \wp'(z))$$

are the affine coordinates of a projective embedding of the torus. Only the point at infinity (corresponding to the origin) is missing from this representation. For more details, look up any book on elliptic functions (e.g. mine).

CHAPTER VII

Homomorphisms and Duality

This chapter describes the elementary theory of homomorphisms and endomorphisms of an abelian manifold. First we relate the rational and complex representations to a purely algebraic representation on the points of finite order. Then we prove the complete reducibility theorem of Poincaré, showing that an abelian manifold admits a product decomposition into simple ones, up to isogeny. Finally, we deal with the duality which arises from the nondegenerate hermitian form, and show how the dual manifold corresponds to divisor classes of divisors algebraically equivalent to 0. The duality includes an essentially algebraic pairing between points and such divisors, and a formula in the last section relates this algebraic pairing with the analytic data and the Riemann form.

§1. The Complex and Rational Representations

Let D, D' be two lattices in V, V' respectively. Any complex analytic homomorphism

$$\lambda_0: V/D \to V'/D'$$

can be lifted to a **C**-linear map $\lambda: V \to V'$ making the following diagram commutative.

$$
\begin{array}{ccc}
V & \xrightarrow{\ \lambda\ } & V' \\
\downarrow & & \downarrow \\
V/D & \xrightarrow{\ \lambda_0\ } & V'/D'
\end{array}
$$

This follows easily by writing a power series expansion for λ_0 locally near the origin, and seeing that the additivity property implies that all terms but those of degree 1 are equal to 0 in this power series. Thus locally near 0, our map λ is C-linear. The global assertion follows, since for any $x \in V$ we can find a large integer N such that x/N is near 0. Then

$$\lambda_0 \left(N \cdot \frac{x}{N} \right) = N \lambda_0 \left(\frac{x}{N} \right)$$

whence $\lambda_0 \equiv \lambda \pmod{D'}$. Of course one can also see this from the general standpoint that the homomorphism λ_0 lifts to the universal covering space V of V/D.

We shall usually use the same symbol λ for the map on V and its induced map on V/D. The ring of complex analytic maps

$$\lambda : V/D \to V/D,$$

i.e., of V/D into itself, will be denoted by $\mathrm{End}(V/D)$. By the above remarks, this ring is represented as a subring of the linear endomorphisms of V, and we call this the **analytic representation of** $\mathrm{End}(V/D)$. Its tensor product with the rational numbers $\mathbf{Q}$ will be denoted by $\mathrm{End}(V/D)_{\mathbf{Q}}$. If $\lambda \in \mathrm{End}(V/D)$ (or $\mathrm{End}(V/D)_{\mathbf{Q}}$), and if we wish to distinguish its induced linear map on V, then we denote the latter by $V(\lambda)$ or λ_V, or $\lambda_{\mathbf{C}}$.

Let V/D be a complex torus as before, and let

$$D = [u_1, \ldots, u_{2n}].$$

We now take $V = \mathbf{C}^n$, that is, we fix a C-basis for V, so that we have complex coordinates for elements of V, and we let $e_1, \ldots, e_n$ be the unit vectors, viewed as column vectors, so that

$$(e_1, \ldots, e_n) = \begin{pmatrix} 1 & 0 & \cdots & 0 \\ 0 & 1 & \cdots & 0 \\ & & \vdots & \\ 0 & 0 & \cdots & 1 \end{pmatrix}$$

Similarly, we view $u_1, \ldots, u_{2n}$ as column vectors, so that $(u_1, \ldots, u_{2n})$ is an $n \times 2n$ matrix of complex numbers.

Let $\lambda \in \mathrm{End}(\mathbf{C}^n/D)$. Then λ has a representation by an $n \times n$ complex matrix $C(\lambda)$ with respect to the basis $\{e_1, \ldots, e_n\}$, and we have

$$(\lambda e_1, \ldots, \lambda e_n) = (e_1, \ldots, e_n) C(\lambda).$$

The matrix multiplication is to be understood formally. That is, if

$$C(\lambda) = (c_{ij}),$$

then

$$(e_1, \ldots, e_n)(c_{ij}) = \left(\ldots, \sum_i c_{ij} e_i, \ldots \right)_{j=1, \ldots, n}.$$

Similar multiplications below are to be understood in the same way. On the other hand, we also have the **rational representation** $Q(\lambda)$, because λ maps D into itself. Thus

$$(\lambda u_1, \ldots, \lambda u_{2n}) = (u_1, \ldots, u_{2n})Q(\lambda),$$

where $Q(\lambda)$ is a $2n \times 2n$ rational matrix. Let U be the $n \times 2n$ matrix consisting of the components of $u_1, \ldots, u_n$, that is

$$U = (u_1, \ldots, u_{2n}).$$

Then

$$(\lambda u_1, \ldots, \lambda u_{2n}) = (e_1, \ldots, e_n) C(\lambda) U$$

and also

$$(\lambda u_1, \ldots, \lambda u_{2n}) = (e_1, \ldots, e_n) U Q(\lambda).$$

Therefore obtain the relation

$$\boxed{C(\lambda)U = UQ(\lambda).}$$

Theorem 1.1. *The matrix*

$$\left(\frac{U}{\overline{U}} \right)$$

is non-singular (where the bar denotes complex conjugate).

Proof. Let M be the above matrix. Suppose M is singular. Then there exists a complex vector Z ($2n$-tuple) such that $MZ = 0$ (we view Z as a column vector). This implies that $UZ = 0$ and $\overline{U}Z = 0$, so that also $U\overline{Z} = 0$.

Hence

$$U(Z + \bar{Z}) = 0 \quad \text{and} \quad U\left(\frac{Z - \bar{Z}}{i}\right) = 0.$$

Both $Z + \bar{Z}$ and $(Z - \bar{Z})/i$ are real matrices, and one of them is not 0. This implies a relation of linear dependence over **R** between the column of U, which is impossible, and proves our assertion.

Theorem 1.2. *The rational representation* $\lambda \mapsto Q(\lambda)$ *is equivalent to the direct sum of the complex representation and its conjugate.*

Proof. By Theorem 1.1, the matrix

$$\begin{pmatrix} U \\ \bar{U} \end{pmatrix}$$

is invertible, and by the above relation, we also get the complex conjugate relation, namely

$$\overline{C(\lambda)}U = \bar{U}Q(\lambda).$$

Therefore

$$\begin{pmatrix} C(\lambda) & 0 \\ 0 & \overline{C(\lambda)} \end{pmatrix}\begin{pmatrix} U \\ \bar{U} \end{pmatrix} = \begin{pmatrix} U \\ \bar{U} \end{pmatrix} Q(\lambda).$$

This proves our theorem.

§2. Rational and p-adic Representations

Let p be a prime number. The points of period p^r on V/D constitute the subgroup

$$(V/D)_r = \frac{1}{p^r} D/D.$$

Let $T_p(V/D)$ be the set of all infinite vectors

$$(a_1, a_2, a_3, \ldots)$$

such that $a_r \in (V/D)_r$ and $pa_{r+1} = a_r$. Then $T_p(V/D)$ is a group under componentwise addition, and is called the **Tate group**. If z is a p-adic integer, we can define an operation of z on $T_p(V/D)$ as follows. We select an ordinary integer m such that $m \equiv z \pmod{p^r}$. If $p^r a = 0$, we define $za = ma$. This is independent of the choice of m. We then define

$$z \cdot (a_1, a_2, \ldots) = (za_1, za_2, \ldots).$$

It is clear that we get an operation of $\mathbf{Z}_p$ on $T_p(V/D)$, which therefore becomes a module over $\mathbf{Z}_p$, and clearly is without torsion.

Actually, $T_p(V/D)$ is free of dimension $2n$ over $\mathbf{Z}_p$, if n is the complex dimension of V.

This is easily proved as follows. Let $x_1, \ldots, x_{2n}$ be vectors in $T_p(V/D)$ whose first components $a_{1,1}, \ldots, a_{2n,1}$ are linearly independent over the field $\mathbf{Z}/p\mathbf{Z}$. Then these vectors are linearly independent over $\mathbf{Z}_p$; for if we had a relation of linear independence, we could assume that not all the coefficients are divisible by p, and hence the projection of this relation on the first component would contradict the hypothesis made on the a_{ij}.

Next, we show that the x_i form a basis of $T_p(V/D)$ over $\mathbf{Z}_p$. We prove this by an inductive argument. Suppose that we can write every element w of $T_p(A)$ as a linear combination

$$(1) \qquad w \equiv z_1 x_1 + \cdots + z_{2n} x_{2n} \pmod{p^r T_p(V/D)},$$

with integers $z_j \in \mathbf{Z}$. Let $w = (b_1, \ldots, b_r, b_{r+1}, \ldots)$. By definition, we have for the first $r + 1$ components

$$z_1(a_{1,1}, \ldots, a_{1,r+1}) + \cdots + z_{2r}(a_{2r,1}, \ldots, a_{2r,r+1})$$
$$= (b_1, \ldots, b_r, b_{r+1}) + (0, \ldots, 0, c_{r+1})$$

for some $c_{r+1} \in (V/D)_{r+1}$. By the very choice of the vectors x_i, there exist integers $d_1, \ldots, d_{2n}$ such that

$$c_{r+1} = d_1 p^r a_{1,r+1} + \cdots + d_{2r} p^r a_{2r,r+1}.$$

If we replace $z_1, \ldots, z_{2r}$ by $z_1 + d_1 p^r, \ldots, z_{2r} + d_{2r} p^r$, we see that we have extended the congruence (1) from r to $r + 1$. This gives us what we wanted.

We define the p-adic counterpart of the space V by letting $V_p(V/D)$ be the set of vectors

$$(a_0, a_1, a_2, \ldots)$$

such that $a_0 \in (V/D)_r$ for some r, and $pa_{r+1} = a_r$ for all $r > 0$. Projecting on the first component gives an exact sequence

$$0 \to T_p(V/D) \to V_p(V/D) \to (V/D)^{(p)} \to 0,$$

where $(V/D)^{(p)}$ is the union of $(V/D)_r$ for all positive integers r, that is, the set of points of p-power order on the torus (V/D).

We have a natural isomorphism

$$\frac{1}{p^r} D/D \approx D/p^r D$$

obtained by multiplication by p^r. Taking the inverse limit over r, we get an isomorphism

(2)
$$\boxed{T_p(V/D) \approx \mathbf{Z}_p \otimes_{\mathbf{Z}} D.}$$

If $\lambda \in \text{End}(V/D)$, then we can represent λ as a $\mathbf{Z}_p$-endomorphism of $T_p(V/D)$ by the obvious action

$$T_p(\lambda)(a_1, a_2, \ldots) = (\lambda a_1, \lambda a_2, \ldots),$$

and, similarly, we get the representation of $V_p(\lambda)$ on $V_p(V/D)$. This representation of $\text{End}(V/D)$ on $V_p(V/D)$ is called the **p-adic representation**.

Theorem 2.1. *The p-adic representation of* $\text{End}(A)$ *is equivalent to the rational representation.*

Proof. Obvious from the boxed isomorphism (2) above.

The p-adic representations on points of finite order were first introduced by Deuring and Weil, apparently more or less simultaneously in 1940, 1941. Deuring used them extensively in his paper "Die Typen der Multiplikatorenringe elliptischer Funktionenkörper," *Abh. Math. Sem. Hamb.* 1941, pp. 197–272, and his previous paper on the theory of correspondences in the same year. Weil mentions them in his paper, "Sur les fonctions algébriques a corps de constantes fini," *C. R. Acad. Sci. Paris* **210** (1940), pp. 592–594, and develops them considerably in his book on abelian varieties. For applications to complex multiplication, cf. the book by Shimura and Taniyama mentioned in the bibliography. Tate notice around 1957 that by taking the inverse limit, i.e., the infinite vectors as defined at the beginning of this section, one could get actually a module over $\mathbf{Z}_p$ (or $\mathbf{Q}_p$), which gives a more natural way to describe the representation.

§3. Homomorphisms

Let $A = V/D$ be a complex torus, and B another complex torus. By Hom(A, B) we mean the additive group of complex analytic homomorphisms of A into B. We shall study some general properties of such homomorphisms.

Let $A = V/D$ be a complex torus as above. By a **subtorus** we mean a torus V'/D', where V' is a complex linear subspace of V, $D' = D \cap V'$, and D' is a lattice in V'.

Let A, B be complex toruses and let

$$f : A \to B$$

be a complex analytic homomorphism. Then Im f *is a complex subtorus of B.*

Proof. Let $A = V/D$ and $B = W/C$, where V, W are finite dimensional vector spaces over the complex numbers, and D, C are lattices. Then f lifts to a **C**-linear map

$$f : V \to W,$$

so $f(V)$ is a **C**-linear subspace of W, and $f(D) \subset C$. Since the image $f(V/D)$ is compact, it follows that $f(D)$ is a lattice in $f(V)$, thus proving that $f(A)$ is a complex subtorus.

Similarly, the kernel of f contains a subtorus as a subgroup of finite index.

Also observe that if $f : A \to B$ is any homomorphism (complex analytic, of course) and $f \neq 0$ then Im f has dimension > 0, and for any integer $N > 0$ we have $Nf \neq 0$. Thus it follows that the natural map

$$\text{Hom}(A, B) \to \text{Hom}(A, B) \otimes \mathbf{Q} = \text{Hom}(A, B)_{\mathbf{Q}}$$

is injectives.

By an **isogeny**

$$f : A \to B$$

we mean a (complex analytic) homomorphism which is surjective and of finite kernel. If f is such an isogeny, and N is a positive integer such that every element of the kernel has period dividing N, then A_N (the group of points of order N) contains Ker f, and consequently there exists a homomorphism

$$g : B \to A$$

such that $g \circ f = N \cdot \text{id}$. Note that $A_N \approx (\mathbf{Z}/N\mathbf{Z})^{2n}$ where $n = \dim A$.

Furthermore, if f is an isogeny, then it has an inverse in $\text{Hom}(B, A)_Q$, namely $N^{-1}g$ where $g \circ f = N \cdot \text{id}$. We denote this inverse in $\text{Hom}(B, A)_Q$ by f^{-1} as usual.

§4. Complete Reducibility of Poincaré

Call a complex torus A **simple** if it has no complex subtorus of dimension > 0. Then any non-zero element $f \in \text{End}(A) = \text{Hom}(A, A)$ must be an isogeny, and consequently has an inverse in $\text{End}(A)_Q$. Therefore $\text{End}(A)_Q$ is a division algebra.

It is false that a subtorus usually has a complementary subtorus. We now use the existence of Riemann forms to prove whatever is true in general, called **Poincaré's complete reducibility** theorem.

Let $A' = V'/D'$ be a subtorus of $A = V/D$. If (V, D) has a Riemann form E, then the restriction of E to (V', D') is obviously a Riemann form, which is non-degenerate if E is non-degenerate. This is clear from the positive definiteness of the associated hermitian form.

Theorem 4.1. *Let $A' = V'/D'$ be a subtorus of $A = V/D$, and assume that (V, D) has a non-degenerate Riemann form. Then there exists a subtorus $A'' = V''/D''$ such that*

$$A = A' + A'' \quad and \quad A' \cap A'' \text{ is finite.}$$

Proof. Let V'' be the orthogonal complement of V' with respect to the positive definite hermitian form H associated with the Riemann form on (V, D). By the Gram-Schmidt orthogonalization process already used in Chapter VI, §3, we can see easily that the orthogonal complement D'' of D' in D (which is discrete) has complementary rank, namely

$$2n - \text{rank of } D'.$$

Hence it is a lattice in V''. The sum $D' + D''$ is then of finite index in D, and the theorem follows. Note that in the orthogonalization process, we can solve first with rational numbers, and then multiply by appropriate positive integers to clear denominators. This gives rise to the finite index.

In Theorem 4.1, it is clear that the sum map

$$A' \times A'' \to A$$

is an isogeny. Theorem 4.1 implies that any abelian manifold is isogenous to a product

$$A_1 \times A_2 \times \cdots \times A_r \to A,$$

such that each factor A_j is **simple**.

The uniqueness of the factors, up to isogeny, then follows according to the usual arguments for semisimple modules in basic algebra.

Similarly, $\text{End}(A) \otimes \mathbf{Q}$ has a direct product decomposition into matrices of endomorphisms. If A is isogenous to a product

$$B \times B \times \cdots \times B$$

of the same abelian manifold repeated r times, then

$$\text{End}(A) \otimes \mathbf{Q} \approx \text{Mat}_r(\text{End}(B) \otimes \mathbf{Q}),$$

where the right hand side is the ring of $r \times r$ matrices, whose components are in $\text{End}(B) \otimes \mathbf{Q}$.

§5. The Dual Abelian Manifold

Again let V be an n-dimensional vector space over $\mathbf{C}$, with a lattice D, and suppose that (V, D) admits a non-degenerate Riemann form. We shall then call the complex analytic torus V/D an **abelian manifold**. When V/D is embedded in a projective space, as an algebraic subvariety, its image in projective space is then called an **abelian variety**. We shall continue, however, to deal with the abelian manifold and theta functions. I follow Weil [7] in this section.

We denote by V^* the complex antidual space of V. It consists of the antifunctionals, that is, of the maps

$$\xi : V \to \mathbf{C}$$

which are **R**-linear and satisfy $\xi(ix) = -i\xi(x)$ for all $x \in V$. We denote the value of ξ at an element x of V by

$$\langle \xi, x \rangle.$$

Thus $(\xi, x) \mapsto \langle \xi, x \rangle$ is a bilinear map from $V^* \times V$ into $\mathbf{C}$.

If $\xi \in V^*$, then $\bar{\xi}$ (whose value at x is $\overline{\xi(x)}$) is a functional, i.e., is **C**-linear. Thus the antifunctionals are merely the complex conjugates of the functionals. Note that V^* is a vector space over $\mathbf{C}$. (We take the antidual space in order to make a certain map φ_E analytic later, instead of anti-analytic.)

If $\xi \in V^*$, then ξ is uniquely determined by its imaginary part. Indeed, write

$$\xi = \rho + i\mu,$$

where $\rho = \text{Re }\xi$ and $\mu = \text{Im }\xi$. Then $\xi(ix) = -i\xi(x)$ means that

$$\mu(x) = \rho(ix) \quad \text{and} \quad \mu(ix) = -\rho(x).$$

Conversely, given an **R**-linear map $\mu : V \to \mathbf{R}$, we can define an antifunctional $\xi = \rho + i\mu$, by letting ρ be the function given in terms of μ by the above relation.

We obtain an **R**-bilinear map

$$(\xi, x) \mapsto \text{Im }\langle \xi, x \rangle,$$

which is *non-degenerate* by the above remark (that μ determines ξ). Therefore, from elementary algebra, we conclude that the set of elements $u^* \in V^*$ such that $\text{Im }\langle u^*, u \rangle \in \mathbf{Z}$ for all $u \in D$, is a lattice D^* in V^*.

Lemma 5.1. *Let* $\mathbf{C}_1$ *be the group of complex numbers of absolute value* 1. *For each* $\xi \in V^*$, *let* χ_ξ *be the element of* $\text{Hom}(D, \mathbf{C}_1)$ *defined by*

$$\chi_\xi(u) = e^{2\pi i \,\text{Im}\,\langle \xi, u \rangle}.$$

Then the map $\xi \mapsto \chi_\xi$ *is an isomorphism between* V^*/D^* *and* $\text{Hom}(D, \mathbf{C}_1)$.

Proof. Our map $\xi \mapsto \chi_\xi$ is clearly a homomorphism. Suppose that

$$\chi_\xi(u) = 1 \quad \text{for all} \quad u \in D.$$

This means that $\xi \in D^*$, and conversely. Hence our map is injective on the factor group V^*/D^*. To see surjectivity, suppose given a homomorphism $D \to \mathbf{R}/\mathbf{Z}$. Since D is free, we can lift this homomorphism to an additive map $\mu : D \to \mathbf{R}$, and such a map can then be extended by **R**-linearity to an **R**-linear map $V \to \mathbf{R}$, since D spans V over **R**. As we remarked above, μ is the imaginary part of a complex functional ξ, thereby proving our lemma.

Let E be a non-degenerate Riemann form on (V, D) and let H be its associated hermitian form, so that E is the imaginary part of H, and

$$H(x, y) = E(ix, y) + iE(x, y).$$

There exists a unique **C**-linear map

$$\varphi_E : V \to V^*$$

such that

$$H(x, y) = \langle \varphi_E(x), y \rangle$$

for all $x, y \subset V$. Taking imaginary parts, and recalling that $E(u, v)$ lies in **Z** for $u, v \in D$, we see that φ_E maps D into D^*, that is, induces a homomorphism, also denoted by φ_E,

$$\varphi_E : V/D \to V^*/D^*.$$

Suppose that E is non-degenerate. Then it is clear that H is non-degenerate and that the map $\varphi_E : V \to V^*$ is an isomorphism. In this case we can **transport** E **to** V^* by defining

$$E^*(\varphi_E(x), \varphi_E(y)) = E(x, y),$$

or equivalently

$$E^*(\xi, \eta) = E(\varphi_E^{-1}(\xi), \varphi_E^{-1}(\eta)).$$

Note that φ_E maps D onto a sublattice of D^*, and hence some integral multiple of φ_E maps a sublattice of D onto D^*. Among other things, we have proved:

Theorem 5.2. *Suppose that E is a non-degenerate Riemann form on (V, D) and let E^* be its transport to V^*. There exists a positive integer m such that mE^* is a non-degenerate Riemann form on (V^*, D^*).*

We now see that if V/D is an abelian manifold, so is V^*/D^*.

Let $\lambda \in \text{End}(V/D)$. We can define as usual its transpose on $\text{End}(V^*/D^*)$ by the condition

$$\langle {}'\lambda\xi, x \rangle = \langle \xi, \lambda x \rangle.$$

From this and a non-degenerate Riemann form E on (V, D) we can define an **involution** on $\text{End}(V/D)$ by the condition

$$\lambda' = \lambda'_E = \varphi_E^{-1}\, {}'\lambda\varphi_E.$$

It will be an immediate consequence of the next theorem that the association $\lambda \mapsto \lambda'$ is indeed an involution, that is an anti-automorphism of period 2.

Theorem 5.3. *Let E be a non-degenerate Riemann form on (V, D), with associated hermitian form H. Let $\lambda' = \lambda'_E$. Then λ' is the adjoint of λ with respect to H, that is*

$$H(\lambda x, y) = H(x, \lambda' y), \qquad x, y \in V.$$

Therefore $\text{tr}(\lambda\lambda') > 0$ *if* $\lambda \neq 0$.

Proof. Trivial computation as follows.

$$H(\lambda'x, y) = H(\varphi_E^{-1} {}^t\lambda\varphi_E(x), y) = \langle {}^t\lambda\varphi_E(x), y \rangle$$
$$= \langle \varphi_E(x), \lambda y \rangle$$
$$= H(x, \lambda y).$$

The assertion about the trace comes from standard elementary linear algebra. Indeed, $\lambda\lambda'$ is self-adjoint, and so can be diagonalized in the representation on V. Furthermore, $\lambda\lambda'$ is positive as an operator on V, so all the diagonal elements are positive, so their sum (the trace) is > 0.

Note that the trace in the theorem is the trace with respect to the complex representation. Since the rational representation is a direct sum of the complex and its conjugate, it follows that the positivity statement also applies to the rational representation. Furthermore, since the rational representation is defined over $\mathbf{Q}$, the trace is $\mathbf{Q}$-valued on $\text{End}(A)$.

§6. Relations with Theta Functions

A normalized theta function is said to be **algebraically equivalent to** 0 if its associated hermitian form H is 0. By (4) of Chapter VI, §1, this means that for such a theta function F, the functional equation takes the shape

$$F(x + u) = F(x)\psi_F(u),$$

where $\psi_F : D \to \mathbf{C}_1$ is a character of D, uniquely determined by F, and which we shall call the **associated character** of F. In previous notation,

$$\psi_F(u) = e^{2\pi i K(u)}.$$

It depends only on the linear equivalence class of F. Conversely, two normalized theta functions which are algebraically equivalent to 0 and have the same associated character are linearly equivalent.

The group of theta functions modulo equivalence, and modulo the subgroup of those linearly equivalent to 0 was defined to be the **Picard group** (of divisor classes). The subgroup of those algebraically equivalent to 0 was denoted by $\text{Pic}_0(V/D)$. We deal only with the latter, and so call it the **Picard group** for short.

The associated character $F \mapsto \psi_F$ above induces an injective homomorphism

$$\text{Pic}_0(V/D) \to \text{Hom}(D, \mathbf{C}_1).$$

We shall see in a moment that it is an isomorphism.

We proved in Chapter VI, Lemma 5.1, that if θ is normalized of type (L, J), then the theta function which is normalized and in the same class as θ_a has type $(L, J - E_a)$. Also recall the map

$$\varphi_\theta : a \mapsto \mathrm{Cl}(\theta_a/\theta),$$

and note that the normalized theta function in the class of θ_a/θ is algebraically equivalent to 0.

The character associated with the linear equivalence class of θ_a/θ is therefore equal to

$$u \mapsto e^{-2\pi i E(a, u)}.$$

Theorem 6.1. *Let θ be an entire theta function with non-degenerate Riemann form E. Then we have a diagram which commutes with factor -1:*

$$
\begin{array}{ccc}
V/D & \xrightarrow{\ \varphi_E\ } & V^*/D^* \\
\varphi_\theta \downarrow & (-1) & \downarrow \chi \\
\mathrm{Pic}_0(V/D) & \xrightarrow[\ \psi\ ]{} & \mathrm{Hom}(D, \mathbf{C}_1)
\end{array}
$$

The right vertical map χ is the isomorphism induced by $\xi \mapsto \chi_\xi$ of Lemma 5.1. The bottom map associates with each element of Pic_0 its associated character, and is an isomorphism. In particular, the homomorphism φ_θ has finite kernel, and $\mathrm{Pic}_0(V/D)$ is isomorphic to V^*/D^*.

Proof. The theorem is obvious by putting together what we know. Start with an element $x \in V$. Its image $\varphi_E(x)$ is characterized by the condition

$$H(x, u) = \langle \varphi_E(x), u \rangle.$$

The corresponding character is given by exponentiating imaginary parts, that is for $\xi = \varphi_E(x)$,

$$\chi_\xi : u \mapsto e^{2\pi i E(x, u)}.$$

On the other hand, we have seen that going around the other way, the associated character of the normalized theta function in the class of θ_x/θ is $e^{-2\pi i E(x, u)}$. This makes the (-1)-commutativity clear.

Since φ_E is surjective, and the right vertical map is an isomorphism, it follows that the bottom map ψ is surjective, so an isomorphism as desired.

Next we identify two possible versions of the transpose of a homomorphism. Let

$$\lambda : V_1/D_1 \to V/D$$

be a homomorphism. For our purposes here, let us denote

$$\lambda^* : V^*/D^* \rightarrow V_1^*/D_1^*$$

the transpose, such that for $z \in V$ and $\xi \in V^*$ we have

$$\langle \lambda^*\xi, z \rangle = \langle \xi, \lambda z \rangle.$$

We use the transpose notation $'\lambda$ for the induced map

$$'\lambda : \text{Pic}_0(V/D) \rightarrow \text{Pic}_0(V_1/D_1)$$

which sends a divisor on V to its inverse image on V_1. In terms of theta functions, if F is a theta function representing a divisor on V/D then $F \circ \lambda$ represents the image of this divisor under $'\lambda$.

Theorem 6.2. *The following diagram is commutative*:

$$
\begin{array}{ccc}
V^*/D^* & \xrightarrow{\ \lambda^*\ } & V_1^*/D_1^* \\
\approx \Big\updownarrow & & \Big\updownarrow \approx \\
\text{Pic}_0(V/D) & \xrightarrow[\ '\lambda\ ]{} & \text{Pic}_0(V_1/D_1).
\end{array}
$$

The vertical maps are the natural isomorphisms, which make correspond

$$\xi \longleftrightarrow \psi_\xi \longleftrightarrow F_\xi$$

where

$$\psi_\xi(u) = e^{2\pi i \text{Im}\langle \xi, u \rangle} \quad and \quad F_\xi(z + u) = F_\xi(z)\psi_\xi(u).$$

Proof. From the equalities

$$F_\xi \circ \lambda(z_1 + u_1) = F_\xi(\lambda z_1)\psi_{F_\xi \circ \lambda}(u_1)$$
$$= F_\xi(\lambda z_1 + \lambda u_1) = F_\xi(\lambda z_1)\psi_{F_\xi}(\lambda u_1)$$

we conclude that

$$\psi_{F_\xi \circ \lambda}(u_1) = \psi_{F_\xi}(\lambda u_1).$$

Then

$$\psi_{\lambda^*\xi}(u_1) = e^{2\pi i \text{Im}\langle \lambda^*\xi, u_1 \rangle} = e^{2\pi i \text{Im}\langle \xi, \lambda u_1 \rangle} = \psi_\xi(\lambda u_1).$$

Therefore $F_{\lambda^*\xi}$ has associated character $\psi_{\lambda^*\xi}$, and the commutativity is clear.

Now let θ be an entire non-degenerate theta function, giving rise to the homormophism

$$\varphi_\theta : V/D \longrightarrow \mathrm{Pic}_0(V/D).$$

Let $\lambda \in \mathrm{End}(V/D)_Q$. We may define an involution

$$\lambda'_\theta = \varphi_\theta^{-1} \circ {}^t\lambda \circ \varphi_\theta$$

where ${}^t\lambda$ is the transpose on the Picard group.

Theorem 6.3. *Let E be the Riemann form associated with θ. Then the two involutions λ'_E and λ'_θ are equal.*

Proof. This is an immediate consequence of the preceding two theorems, and of the diagram:

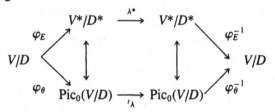

The central square is commutative, and the two end triangles have sign -1, which cancels.

Let X be a divisor on V/D. We denote by $\mathscr{C}(X)$ the set of all divisors Y on V/D for which there exist two positive integers m, m' such that mX is algebraically equivalent to Y. We call $\mathscr{C}(X)$ a **polarization** of V/D if $\mathscr{C}(X)$ contains the positive divisor of an entire non-degenerate theta function; or equivalently, if some positive multiple of a divisor is a hyperplane section in some projective embedding. Suppose that X is the divisor of the entire theta function θ, and is non-degenerate. Then it is clear that the involution

$$\lambda \mapsto \lambda'_\theta$$

depends only on the class $\mathscr{C}(X)$, and therefore λ'_θ is also denoted by $\lambda'_\mathscr{C}$, where $\mathscr{C}$ is the polarization $\mathscr{C} = \mathscr{C}(X)$. Theorem 6.3 gives an analytic description of the involution defined purely algebraically by the formula

$$\lambda \mapsto \lambda'_\mathscr{C}.$$

§7. The Kummer Pairing

We shall now describe the p-adic version of the duality between V/D and its dual V^*/D^*.

Let $\xi \in V^*$ be such that $N\xi \in D^*$ for some positive integer N. (In the application later, we shall consider the case when N is a power of p, but for the moment, it is notationally convenient to work with arbitrary integers.) We recall the two isomorphisms

$$\text{Pic}_0(V/D) \approx \text{Hom } (D, \mathbf{C}_1) \approx V^*/D^*.$$

Therefore, with ξ we can associate a normalized theta function (of course not entire) f_ξ, well defined up to multiplication by an abelian function, algebraically equivalent to 0, and satisfying

$$f_\xi(x + u) = f(x)\psi_\xi(u) = f(x)e^{2\pi i \text{Im}\langle \xi, u \rangle},$$

for $x \in V$ and $u \in D$. Given a finite number of values of $u \in D$, we can always find x sufficiently general in V such that the terms in the above equation are defined. The condition $N\xi \in D^*$ implies that f_ξ^N is an abelian function. If $a \in V$ and $Na \in D$, it follows that there exists an N-th root of unity $e_N(\xi, a)$ such that

$$\boxed{f_\xi(Nx + Na) = e_N(\xi, a)f_\xi(Nx).}$$

It is clear that this root of unity does not depend on the choice of f_ξ in a linear equivalent class, nor on the choice of $\xi \pmod{D^*}$ and $a \pmod{D}$. Therefore the association

$$(\xi, a) \mapsto e_N(\xi, a)$$

gives a well-defined map

$$(V^*/D^*)_N \times (V/D)_N \to \mu_N,$$

where μ_N is the group of N-th roots of unity, and it is trivially verified that this map is $\mathbf{Z}$-bilinear, i.e. is a pairing. It is equally easily verified that the kernels on both sides are 0, i.e. that the map puts $(V^*/D^*)_N$ and $(V/D)_N$ in perfect duality.

There is a special case of ξ which is of interest. Suppose that θ is a nondegenerate entire theta function, with associated Riemann form E. If $b \in V$, then there is a unique normalized theta function f in $\text{Cl}(\theta_b/\theta)$, whose associated character is $-E_b$. We suppose that $Nb \equiv 0 \pmod{D}$, and $Na \equiv 0 \pmod{D}$, in other words, b and a represent points of order N in V/D. Then the root of unity e_N corresponding to this special choice can be written in the form

$$e_N(\varphi_E(b),\, a) = e^{-2\pi i N E(b,a)}$$

as is obvious from the definitions, and Theorem 6.1.

Now let us pass to the p-adic version. Instead of N we use p^N in the preceding discussion. Then $(V/D)_N$ really means $(V/D)_{p^N}$, and similarly for $(V^*/D^*)_N$ and μ_N. When we pass from p^N to p^{N+1}, it is easily seen that the pairing is consistent with respect to the inverse limit, that is, that if

$$p^{N+1}\xi \in D^* \quad \text{and} \quad p^{N+1}a \in D,$$

then

$$e_{N+1}(\xi,\, a)^p = e_N(p\xi,\, pa).$$

Therefore we can express our pairing in terms of the Tate group, and we obtain a $\mathbf{Z}_p$-bilinear map

$$T_p(V^*/D^*) \times T_p(V/D) \to T_p(\mu),$$

which make the two $\mathbf{Z}_p$-modules on the left $\mathbf{Z}_p$-dual to each other. Note that $T_p(\mu)$ is formed of vectors

$$(\zeta_1,\, \zeta_2,\, \zeta_3,\, \ldots)$$

where ζ_N is a p^N-th root of unity such that $\zeta_{N+1}^p = \zeta_N$. Thus $T_p(\mu)$ is a 1-dimensional space over $\mathbf{Z}_p$, which cannot be identified with $\mathbf{Z}_p$ unless we have selected a basis.

Remark. The advantage of the root of unity $e_N(\xi, a)$ and the p-adic pairing is that they hold *algebraically*. When we embed an abelian torus in projective space, we obtain the complex points $A_{\mathbf{C}}$ of an abelien variety A, which may be defined by equations over a field k. As we shall see in the next chapter, the zeros of a theta function in V correspond to a divisor on A. One can then see that the definition of $e_N(\xi, a)$ corresponds to a purely algebraic definition involving only the divisor class corresponding to ξ on A. The resulting roots of unity then lie in the complex numbers, containing k, but generate possibly extensions of k (this is always the case if we take k to be finitely generated over $\mathbf{Q}$, and N sufficiently large). Then these roots of unity, and $T_p(\mu)$, have the definite advantage that they can be used as representation spaces for the Galois group of the algebraic closure of k, thus giving an additional tool which takes essential account of more arithmetic aspects of the situation. It is then absurd in such a context to identify $T_p(\mu)$ with $\mathbf{Z}_p$, since a Galois group operates trivially on $\mathbf{Z}_p$.

For the algebraic theory and applications, the reader is referred to the books of Lang, Mumford, Shimura, and Weil.

§8. Periods and Homology

Let $A = V/D$ be a complex torus, and pick an isomorphism of V with $\mathbf{C}^n$, with complex coordinates $z_1, \ldots, z_n$.

We first observe that the holomorphic 1-forms on A are spanned over the complex numbers by

$$dz_1, \ldots, dz_n.$$

Proof. The coordinates $z_1, \ldots, z_n$ define a complex analytic chart at every point of A. Any holomorphic 1-form has an expression

$$f_1(z)dz_1 + \cdots + f_n(z)dz_n$$

at every point, where the functions $f_1, \ldots, f_n$ are holomorphic, and the functions which form the coefficient of dz_1 at every point are then analytic continuations of each other on the torus. Hence they are holomorphic on the torus, i.e. periodic, so constant. Hence $dz_1, \ldots, dz_n$ form a basis of the 1-forms over $\mathbf{C}$, as was to be shown.

Denote the vector space of holomorphic 1-forms on A by $\Omega_1(A) = \Omega_1$. We have a pairing

$$H_1(A, \mathbf{Z}) \times \Omega_1 \to \mathbf{C}$$

given by

$$(\gamma, \omega) \mapsto \int_\gamma \omega.$$

This induces a homormophism

$$H_1(A, \mathbf{Z}) \to \hat{\Omega}_1$$

into the dual space. We shall prove that this map is injective, and that its image is a lattice, giving rise to an isomorphism of $\hat{\Omega}_1/H_1(A, \mathbf{Z})$ with A itself.

First we note that Ω_1 can be viewed as a real vector space, and that it has dimension $2n$ over $\mathbf{R}$. Indeed, the forms

$$dz_1, \ldots, dz_n, idz_1, \ldots, idz_n$$

are linearly independent over **R**, as one sees immediately since if we write $z_k = x_k + iy_k$, then

$$dz_1, dy_1, \ldots, dx_n, dy_n$$

are independent real coordinates.

If ω is perpendicular to every cycle in $H_1(A, \mathbf{Z})$, then the map

$$P \mapsto \int_0^P \omega$$

defines a holomorphic function on A, which is therefore constant, and therefore equal to 0. Hence $\omega = 0$. This implies that in the above pairing, the kernel in Ω_1 is 0. Since $H_1(A, \mathbf{Z})$ is a free abelian group on $2n$ generators, and therefore its image in $\hat{\Omega}_1$ has real rank $\leq 2n$, it follows that this real rank must be equal to $2n$, and therefore that $H_1(A, \mathbf{Z})$ may be viewed as a lattice in $\hat{\Omega}_1$.

There is also a natural isomorphism

$$D \xrightarrow{\sim} H_1(A, \mathbf{Z}),$$

obtained as follows. The group D may be viewed as the fundamental group of the covering

$$p: V \to V/D = A.$$

If $u \in D$ and l_u is a path from 0 to u in V, then $p \circ l_u$ is a cycle on A. The association

$$u \mapsto \text{class of } p \circ l_u \text{ in } H_1(A, \mathbf{Z})$$

gives rise to the above isomorphism.

Thus we obtain a natural embedding

$$\rho_D: D \xrightarrow{\sim} H_1(A, \mathbf{Z}) \to \hat{\Omega}_1,$$

whose image in $\hat{\Omega}_1$ will be denoted by $\hat{D}$.

On the other hand, let $z \in V$. Let l_z be any path from 0 to z. Then $p \circ l_z$ is a path in A, which we may view as an element of $\hat{\Omega}_1$, that is, giving rise to the functional

$$\omega \mapsto \int_{p \circ l_z} \omega = \int_0^z p^*\omega.$$

Thus we have extended the map ρ to a map

$$\rho_V : V \to \hat{\Omega}_1.$$

Theorem 8.1. *The map ρ_V is a $\mathbf{C}$-linear isomorphism, inducing an isomorphism*

$$\rho_A : V/D \to \hat{\Omega}_1/\hat{D} = \hat{\Omega}_1/H_1(A, \mathbf{Z}).$$

Proof. For each $j = 1, \ldots, n$ we have

$$\int_0^z dz_j = z_j.$$

Thus it is clear that the map is $\mathbf{C}$-linear, and we have already seen above that it induces an injective map on D. This proves the theorem, since D generates V over $\mathbf{R}$, and its image generates $\hat{\Omega}_1$ over $\mathbf{R}$ (being a lattice).

Now let $\{\omega_1, \ldots, \omega_n\}$ be a basis of Ω_1 over $\mathbf{C}$. Let Λ be the abelian group of periods, that is the group of elements

$$\int_\gamma (\omega_1, \ldots, \omega_n), \text{ for all } \gamma \in H_1(A, \mathbf{Z}).$$

Theorem 8.2. *The map*

$$P \mapsto \int_0^P (\omega_1, \ldots, \omega_n)$$

establishes a complex analytic isomorphism of $A = V/D$ with $\mathbf{C}^n/\Lambda$, sending D on Λ.

Proof. The map is well defined, from V into $\mathbf{C}^n/\Lambda$. The above duality immediately implies that the kernel is D. The map is locally surjective in a neighborhood of the origin because $dz_1, \ldots, dz_n$ are local analytic coordinates, and it is surjective by additivity, and the fact that given $z \in \mathbf{C}^n$, the element z/N is close to the origin for N a large positive integer. Since the map is obviously analytic, the theorem follows.

Theorem 8.3. *Let*

$$f : V/D \to V'/D'$$

be a complex analytic map such that $f(0) = 0$. Then f is a homomorphism.

Proof. Let $\Omega_1(f): \Omega_1(V'/D') \to \Omega_1(V/D)$ be the induced linear map on holomorphic differential forms, obtained by pull back. We further get the dual map (linear)

$$\hat{\Omega}_1(f) : \hat{\Omega}_1(V/D) \to \hat{\Omega}_1(V'/D'),$$

and it is trivially verified by the change of variables formula that we have a commutative diagram:

$$
\begin{array}{ccc}
\hat{\Omega}_1(V/D) & \xrightarrow{\hat{\Omega}_1(f)} & \hat{\Omega}_1(V'/D') \\
\rho_V \uparrow \approx & & \approx \uparrow \rho_{V'} \\
V & \xrightarrow{f} & V'
\end{array}
$$

Hence f is linear, as was to be shown.

CHAPTER VIII

Riemann Matrices and Classical Theta Functions

§1. Riemann Matrices

Let H be a positive definite hermitian form on $\mathbf{C}^n$. We may write H in terms of its real and imaginary parts as

$$H(u, v) = E(iu, v) + iE(u, v)$$

where E is alternating and real valued. By convention, H is linear in the first variable, and anti-linear in the second. Let Λ be a lattice in $\mathbf{C}^n$. We call H (or E) a **Riemann form for the pair** $(\mathbf{C}^n, \Lambda)$ if

$$E(u, v) \in \mathbf{Z} \quad \text{for} \quad u, v \in \Lambda.$$

Note that the form $(u, v) \mapsto E(iu, v)$ is symmetric positive definite. Let

$$W = (w_1, \ldots, w_{2n})$$

be a basis for Λ over $\mathbf{R}$. We view w_j $(j = 1, \ldots, 2n)$ as column vectors in $\mathbf{C}^n$, so that W is an $n \times 2n$ matrix. Elements of $\mathbf{C}^n$ may then be written in the form

$$Wx \quad \text{with} \quad x \in \mathbf{R}^{2n},$$

and elements of the lattice may be written in this form with $x \in \mathbf{Z}^{2n}$. There is a unique alternating matrix P such that

$$E(Wx, \ Wy) = {}^t x P y.$$

"Alternating" means that ${}^t P = -P$. Thus the matrix P is the matrix representing the alternating form with respect to the standard basis of $\mathbf{R}^{2n}$. In fact, if $P = (p_{ij})$ then

$$E(w_i, \ w_j) = p_{ij}, \quad i, j = 1, \ldots, 2n.$$

Let C be the $2n \times 2n$ real matrix such that

$$iW = WC.$$

Then

$$E(iWx, \ Wy) = E(WCx, \ Wy) = {}^t x \, {}^t CPy.$$

It follows that ${}^t CP$ is symmetric positive, and we also have

$$ {}^t CP = -PC.$$

The following two lemmas express in terms of the relevant matrices two conditions characterizing a Riemann form, namely the facts that the form $E(iWx, \ Wy)$ is symmetric and positive definite.

Lemma 1.1. *The symmetry of* ${}^t CP$ *is equivalent with the condition*

$$WP^{-1} \, {}^t W = 0.$$

Proof. This symmetry is equivalent with the condition

$$- \left(\frac{W}{\overline{W}} \right) CP^{-1}({}^t W, \ {}^t \overline{W}) = \left(\frac{W}{\overline{W}} \right) P^{-1} \, {}^t C({}^t W, \ {}^t \overline{W}).$$

Using once more the definition $iW = WC$ and performing the matrix multiplications, yields the desired conclusion.

Lemma 1.2. *The matrix associated with H is*

$$M = 2i(\overline{W}P^{-1} \, {}^t W)^{-1} > 0,$$

and so for $u, v \in \mathbf{C}^n$ we have

$$H(u, \ v) = {}^t u M \overline{v}.$$

Proof. Let M be the matrix as indicated above. It is clear that M is hermitian. Thus it will suffice to show that

$$H(u, u) = {}^t u M \bar{u}.$$

Let $i_n = i 1_n$ be i times the unit $n \times n$ matrix. Then

$$\begin{pmatrix} i_n & 0 \\ 0 & -i_n \end{pmatrix} \begin{pmatrix} W \\ \overline{W} \end{pmatrix} = \begin{pmatrix} W \\ \overline{W} \end{pmatrix} C.$$

Since $H(u, u) = E(iu, u)$, we put $u = Wx$ to obtain:

$$H(u, u) = -{}^t x P C x = -{}^t x P \begin{pmatrix} W \\ \overline{W} \end{pmatrix}^{-1} \begin{pmatrix} i_n & 0 \\ 0 & -i_n \end{pmatrix} \begin{pmatrix} W \\ \overline{W} \end{pmatrix} x$$

$$= -{}^t x ({}^t W, {}^t \overline{W})({}^t W, {}^t \overline{W})^{-1} P \begin{pmatrix} W \\ \overline{W} \end{pmatrix}^{-1} \begin{pmatrix} i_n & 0 \\ 0 & -i_n \end{pmatrix} \begin{pmatrix} W \\ \overline{W} \end{pmatrix} x.$$

The terms on the far left and far right are $({}^t u, {}^t \bar{u})$ and $(u, \bar{u})$ respectively. We then carry out the matrix multiplication for the middle product which is equal to

$$\left[\begin{pmatrix} W \\ \overline{W} \end{pmatrix} P^{-1} ({}^t W, {}^t \overline{W}) \right]^{-1}.$$

We use Lemma 1.1, and we use the fact that

$$\begin{pmatrix} 0 & X \\ Y & 0 \end{pmatrix}^{-1} = \begin{pmatrix} 0 & Y^{-1} \\ X^{-1} & 0 \end{pmatrix}$$

with the appropriate matrices X, Y to find that the above expression for $H(u, u)$ is equal to

$$i\,{}^t u (\overline{W} P^{-1}\,{}^t W)^{-1} \bar{u} - i\,{}^t \bar{u} (W P^{-1}\,{}^t \overline{W})^{-1} u = 2i\,{}^t u (\overline{W} P^{-1}\,{}^t W)^{-1} \bar{u},$$

This concludes the proof.

Now let $W = (\omega_1, \omega_2)$ where ω_1, ω_2 are square matrices, let

$$J = \begin{pmatrix} 0 & 1_n \\ -1_n & 0 \end{pmatrix}$$

and suppose that $P = J$. Performing the matrix multiplication, we see that relations of Lemmas 1.1 and 1.2 can be rewritten as:

RR 1. $\omega_2 \,{}^t \omega_1 - \omega_1 \,{}^t \omega_2 = 0$, *or equivalently* $WJ\,{}^t W = 0$.

RR 2. $i(\overline{\omega}_2{}'\omega_1 - \overline{\omega}_1{}'\omega_2)^{-1} > 0$, *or equivalently* $i\overline{W}J{}'W > 0$.

These are called the **Riemann relations**. Taking an inverse and a complex conjugate, we see that the second one is equivalent with

RR 2'. $i(\omega_2{}'\overline{\omega}_1 - \omega_1{}'\overline{\omega}_2) > 0$.

Let $\mathfrak{R}$ be the set of all pairs (ω_1, ω_2) of $n \times n$ complex matrices satisfying the Riemann relations above. As usual, we let $\mathrm{Sp}_{2n}(\mathbf{R})$ be the subgroup of $\mathrm{GL}_{2n}(\mathbf{R})$ consisting of all matrices M such that

$$MJ{}'M = J.$$

Then Sp_{2n} is closed under transpose.

Let $\mathfrak{H}_n$ be the **Siegel upper half space**, consisting of all matrices

$$z \in \mathrm{Mat}_n(\mathbf{C}) \ (n \times n \text{ complex matrices})$$

which are symmetric and whose imaginary part is symmetric positive definite.

Lemma 1.3. *Let* $(\omega_1, \omega_2) \in \mathfrak{R}$.

(i) *If* $g \in \mathrm{GL}_n(\mathbf{C})$ *then* $g(\omega_1, \omega_2) \in \mathfrak{R}$.

(ii) *If* $M \in \mathrm{Sp}_{2n}(\mathbf{R})$ *then* $(\omega_1, \omega_2)M \in \mathfrak{R}$.

(iii) *The matrices* ω_1, ω_2 *are invertible.*

(iv) *We have* $\omega_2^{-1}\omega_1 \in \mathfrak{H}_n$.

Proof. The first two assertions are immediate from the definitions of $\mathfrak{R}$, Sp_{2n}, and the Riemann relations. As for (iii), suppose there exists a vector $v \in \mathbf{C}^n$ such that $'v\omega_1 = 0$. Then

$$'v(\omega_1{}'\overline{\omega}_2 - \omega_2{}'\overline{\omega}_1)\overline{v} = 0,$$

whence $v = 0$ by **RR 2**. Hence ω_1^{-1} exists. Furthermore $J \in \mathrm{Sp}_{2n}(\mathbf{R})$, so

$$(-\omega_2, \omega_1) = (\omega_1, \omega_2)J \in \mathfrak{R}$$

by (ii), and therefore ω_2^{-1} exists, thus proving (iii). Finally $(\omega_2^{-1}\omega_1, 1_n) \in \mathfrak{R}$ by (i) and therefore (iv) follows from the Riemann relations **RR 2**. This concludes the proof.

In light of Lemma 1.3, if $\omega_1 = z$ and $\omega_2 = 1_n$, then the second Riemann Relations read:

RR 2''. $i(z - \overline{z})^{-1} > 0$.

We have to consider the slightly more general case when the matrix associated with the Riemann form E is not J but has elementary divisors. Thus let $d_1, \ldots, d_n$ be positive integers such that

$$d_1 \mid d_2 \mid \cdots \mid d_n,$$

and put

$$\delta = \begin{pmatrix} d_1 & & & \\ & d_2 & & O \\ & & \ddots & \\ O & & & d_n \end{pmatrix} = \operatorname{diag}(d_1, \ldots, d_n).$$

Given the Riemann form E integral valued on the lattice Λ, there exists a basis $\{w_1', \ldots, w_{2n}'\}$ of Λ such that

$$E(w_j', w_k') = J_\delta = \begin{pmatrix} 0 & \delta \\ -\delta & 0 \end{pmatrix}.$$

Then

$$E(W'x, W'y) = {}^t x J_\delta y \quad \text{for} \quad x, y \in \mathbf{R}^{2n}.$$

Let

$$U = \begin{pmatrix} 1 & 0 \\ 0 & \delta \end{pmatrix} \quad \text{and} \quad W = W' U^{-1} \quad \text{so} \quad W' = WU.$$

Then matrix multiplication shows that

$$E(Wx, Wy) = {}^t x J y$$

and

$$\Lambda = W' \mathbf{Z}^{2n} = WU \mathbf{Z}^{2n} = (\omega_1, \omega_2 \delta) \mathbf{Z}^{2n}.$$

This kind of changes of coordinates reduces the situation to the preceding one, when all the elementary divisors are equal to 1.

§2. The Siegel Upper Half Space

A **polarization** of a complex torus (for this section) is a choice of Riemann form. Thus a polarized complex torus is a triple (V, Λ, E) consisting of a complex n-dimensional space V, a lattice Λ, and a Riemann form E. We say that this is a **principal polarization** if the elementary divisors $d_1, \ldots, d_n$ of E with respect to a Frobenius basis are all equal to 1. We shall now study this case for simplicity of notation, and obtain a classification of such abelian manifolds up to isomorphisms, by parametrizing them as a quotient of a generalized (Siegel) upper half plane modulo the action of a group of automorphisms.

If we have chosen an isomorphism of V with $\mathbf{C}^n$, and a $\mathbf{Z}$-basis for the lattice Λ, so that its elements can be written as an $n \times 2n$ matrix W, and the form E is represented by the matrix P with respect to this basis, then we also write a triple

$$(V, \Lambda, E) = (\mathbf{C}^n, W, P).$$

Suppose that the matrix P representing E is of the form

$$J = \begin{pmatrix} O & 1_n \\ -1_n & O \end{pmatrix},$$

where 1_n is the identity $n \times n$ matrix. This corresponds to having chosen a Frobenius basis for the lattice, and to all the elementary divisors $d_1, \ldots, d_n$ being equal to 1.

To each matrix z in $\mathfrak{H}_n$ we may associate the corresponding $n \times 2n$ matrix $(z, 1)$ which we may view as a matrix W of the components of a basis for the lattice as at the beginning of the preceding section.

Lemma 2.1. *Every isomorphism class of principally polarized abelian manifold contains a representative*

$$(\mathbf{C}^n, (z, 1_n), J) \quad with \quad z \in \mathfrak{H}_n,$$

for which the columns of $(z, 1_n)$ *form a Frobenius basis.*

Proof. Let $W = (\omega_1, \omega_2)$ be a matrix satisfying the Riemann relations, and whose columns form a basis for the period lattice of the abelian manifold. We note:

If $\omega_2 = 1_n$, *then the Riemann relations are equivalent with the property that* $\omega_1 \in \mathfrak{H}_n$.

Since multiplication by ω_2^{-1} on $\mathbf{C}^n$ is a linear isomorphism, we see that Lemma 2.1 follows immediately from Lemma 1.3.

To get uniqueness we have to factor out by an appropriate group of automorphisms. As before, let:

$\mathfrak{R}$ = set of all pairs (ω_1, ω_2) satisfying the Riemann relations;

$Sp_{2n}(\mathbf{R})$ = subgroup of $GL_{2n}(\mathbf{R})$ consisting of all matrices γ such that

$$\gamma J' \gamma = J.$$

Then $Sp_{2n}(\mathbf{R})$ is closed under transpose. Write

$$\gamma = \begin{pmatrix} a & b \\ c & d \end{pmatrix}.$$

We have:

$\gamma J' \gamma = J$ *if and only if* $a'd - b'c = 1_n$ *and* $a'b, c'd$ *are symmetric.*

When $n = 1$, this means that γ lies in SL_2.

We shall see that the association

$$z \mapsto (az + b)(cz + d)^{-1} = \gamma(z)$$

defines an operation of $Sp_{2n}(\mathbf{R})$ on $\mathfrak{H}_n$. We have by Lemma 1.3 (ii)

$$(z, 1_n)\begin{pmatrix} a & b \\ c & d \end{pmatrix} = (za + c, zb + d) \in \mathfrak{R}.$$

Hence by (iv).

$$(zb + d)^{-1}(za + c) \in \mathfrak{H}_n.$$

But $'z = z$ so taking transposes yields

$$('az + 'c)('bz + 'd)^{-1} \in \mathfrak{H}_n.$$

Since $Sp_{2n}(\mathbf{R})$ is closed under transposes, it follows that $\gamma(z) \in \mathfrak{H}_n$. It is then immediately verified that this defines an operation of $Sp_{2n}(\mathbf{R})$ on $\mathfrak{H}_n$.

Suppose that (ω_1', ω_2') is another basis for the lattice Λ. Then there is a matrix $M \in GL_{2n}(\mathbf{Z})$ such that

$$(\omega_1', \omega_2') = (\omega_1, \omega_2)'M,$$

and conversely. The form E with matrix J with respect to the basis (ω_1, ω_2) has the matrix

$$MJ'M$$

with respect to the basis (ω_1', ω_2'). Therefore the new basis has the same matrix for the Riemann form if and only if $M \in Sp_{2n}$. We let

$$\Gamma = \mathrm{Aut}(J) = \mathrm{GL}_{2n}(\mathbf{Z}) \cap \mathrm{Sp}_{2n}(\mathbf{R}) = \mathrm{Sp}_{2n}(\mathbf{Z}).$$

When $n = 1$ then $\Gamma = \mathrm{SL}_2(\mathbf{Z})$.

We have now seen how the choices of a basis for the lattice, and a linear isomorphism of $\mathbf{C}^n$ affect the choices we have made. We define $\mathscr{A}$ to be the set of equivalence class of polarized abelian manifolds. Any two representatives

$$(\mathbf{C}^n, (z, 1_n), J) \quad \text{and} \quad (\mathbf{C}^n, (z', 1_n), J)$$

differ by a change of basis of the lattice preserving the matrix J as above, followed by a linear isomorphism on $\mathbf{C}^n$ (multiplication by an invertible matrix on the left) transforming the second half of the lattice vectors into the unit vectors. Hence we have proved:

Theorem 2.2. *The map*

$$z \mapsto \text{isomorphism class of } (\mathbf{C}^n, (z, 1_n), J)$$

induces a bijection

$$\Gamma \backslash \mathfrak{H}_n \to \mathscr{A}.$$

§3. Fundamental Theta Functions

This section follows Shimura [Sh].

Let Λ be a lattice in $\mathbf{C}^n$. We recall that a theta function on $\mathbf{C}^n$ with respect to Λ is an entire function f satisfying the condition

$$f(u + l) = f(u)\mathbf{e}(\lambda(u, l))\psi(l)$$

for $u \in \mathbf{C}^n$, $l \in \Lambda$, and $\mathbf{e}(z) = e^{2\pi i z}$. Here ψ is an arbitrary function, and λ is $\mathbf{C}$-linear in u, $\mathbf{R}$-linear in l. A theta function is called **normalized** if there is a hermitian positive form H such that

$$f(u + l) = f(u)\mathbf{e}\left(\frac{1}{2i}H\left(u + \frac{l}{2}, l\right)\right)\psi(l)$$

and if ψ has absolute value equal to 1. We let E be the imaginary part of H, so E is real valued and alternating. We assume that H is $\mathbf{C}$-linear in its first variable, anti-linear in its second variable. It follows from the above assumptions that

$$\psi(l + l') = \psi(l)\psi(l')\mathbf{e}\left(\frac{1}{2} E(l, l')\right).$$

We denote by $\text{Th}(H, \psi, \Lambda)$ the vector space of normalized theta functions satisfying the above conditions with H, ψ, and call (H, ψ) a **type** for such theta functions. This notation is better adapted to the applications we now have in mind than the previous notation in Chapter VI.

Let $z \in \mathfrak{H}_n$ and for $r, s \in \mathbf{R}^n$ define

$$\theta(u, z, r, s) = \sum_{m \in \mathbf{Z}^n} \mathbf{e}\left(\frac{1}{2}\,{}^t(r + m)z(r + m) + {}^t(r + m)(u + s)\right).$$

The positive definiteness of the imaginary part of z insures that the exponential term goes to zero like e^{-cm^2} for some $c > 0$, whence we have absolute convergence, uniform for u in any compact set of $\mathbf{C}^n$. We shall now state three simple transformation formulas.

Th 1. *For $a, b \in \mathbf{R}^n$ we have*

$$\theta(u, z, r + a, b + s)$$
$$= \mathbf{e}\left(\frac{1}{2}\,{}^t aza + {}^t a(u + s + b)\right)\theta(u + za + b, z, r, s).$$

This follows directly from the above definition, expanding out the exponent and collecting terms.

Th 2. *For $a, b \in \mathbf{Z}^n$ we have*

$$\theta(u, z, r + a, s + b) = \mathbf{e}({}^t rb)\,\theta(u, z, r, s).$$

Again this follows directly by observing that the sum expressing the left hand side can first be changed by replacing $m + a$ by m in the sum over m, and then the simple term $\mathbf{e}({}^t rb)$ comes out as a factor. Combining these two formulas yields:

Th 3. *If $a, b \in \mathbf{Z}^n$ then*

$$\theta(u + za + b, z, r, s) = \mathbf{e}\left(-\frac{1}{2}\,{}^t aza - {}^t au + {}^t rb - {}^t sa\right)\theta(u, z, r, s).$$

$$= \mathbf{e}\left(-\frac{1}{2}\,{}^t aza - {}^t au\right)\chi_{-s,r}(a, b)\theta(u, z, r, s)$$

where $\chi_{-s,r}(a, b) = \mathbf{e}(-{}^t sa + {}^t rb)$ is a character.

Note that the theta function $\theta(u, z, r, s)$ (as function of u) is not normalized. We shall multiply it by a trivial theta function so as to obtain a normalized one. We let

$$Q(u, z) = \frac{1}{2} {}^{t}u(z - \bar{z})^{-1}u$$

so $Q(u, z)$ is quadratic in u. Then $\mathbf{e}(Q(u, z))$ is a trivial theta function, and

$$Q(u + l) - Q(u, l) = {}^{t}\left(u + \frac{l}{2}\right)(z - \bar{z})^{-1}l.$$

Note that $Q(u, z)$ is not holomorphic in z. Define as in Shimura

$$\varphi(u, z, r, s) = \mathbf{e}(Q(u, z))\,\theta(u, z, r, s).$$

We shall find again the hermitian form of the last section, defined by

$$H(u, v) = 2i\,{}^{t}u(z - \bar{z})^{-1}\bar{v}.$$

Note that putting $l = za + b$ with $a, b \in \mathbf{Z}^n$ we find

$$\frac{1}{2i} H\left(u + \frac{l}{2}, l\right) = {}^{t}\left(u + \frac{l}{2}\right)a + {}^{t}\left(u + \frac{l}{2}\right)(z - \bar{z})^{-1}l.$$

Th 4. *For $a, b \in \mathbf{Z}^n$ we have*

$$\frac{\varphi(u + az + b, z, r, s)}{\varphi(u, z, r, s)} = \mathbf{e}\left(\frac{1}{2i} H\left(u + \frac{l}{2}, l\right)\right)\mathbf{e}\left(\frac{1}{2} {}^{t}ab\right)\chi_{-s,r}(a, b).$$

This suggests that we define

$$\psi_z(za + b) = \mathbf{e}\left(\frac{1}{2} {}^{t}ab\right) \quad \text{and} \quad \psi_{z,r,s}(za + b) = \mathbf{e}\left(\frac{1}{2} {}^{t}ab + {}^{t}rb - {}^{t}sa\right).$$

Thus

$$\psi_{z,r,s} = \psi_z \chi_{z,-s,r}$$

where $\chi_{z,-s,r}$ is the character defined by

$$\chi_{z,-s,r}(za + b) = \mathbf{e}({}^{t}rb - {}^{t}sa).$$

Let $[z, 1_n]$ be the lattice of all points $za + b$ with $a, b \in \mathbf{Z}^n$. Then with the above definitions, we find that **Th 4** can be stated as follows.

The functions $\varphi(u, z, r, s)$ *lie in* $\mathrm{Th}(H_z, \psi_z X_{z,-s,r}, [z, 1_n])$.

Now let X be any character on $\mathbf{C}^n$, viewed as $\mathbf{R}^{2n}$ if we express elements in the form

$$za + b \quad \text{with} \quad a, b \in \mathbf{R}^n.$$

We observe that the map

$$\theta \mapsto \theta X$$

gives an isomorphism

$$\mathrm{Th}(H, \psi, \Lambda) \xrightarrow{\approx} \mathrm{Th}(H, \psi X, \Lambda).$$

On the other hand, define

$$\mathrm{Th}(z) = \text{space of entire functions satisfying}$$

$$\frac{f(u + za + b)}{f(u)} = \mathbf{e}\left(-\frac{1}{2}\,{}^t a z a - {}^t a u\right), \quad \text{for } \cdot a, b \in \mathbf{Z}^n.$$

Then certainly the functions $\theta(u, z, r, s)X_{z,-s,r}(u)$ lie in this space. Thus we have isomorphisms

$$\mathrm{Th}(z) \approx \mathrm{Th}(H_z, \psi_z, [z, 1_n]) \approx \mathrm{Th}(H_z, \psi_{z,r,s}, [z, 1_n]).$$

Each one is obtained by multiplication with an appropriate function.

Theorem 3.1. *Let* $L = [z, \delta]$, *where* $\delta = \mathrm{diag}(d_1, \ldots, d_n)$ *with*

$$0 < d_1 \big| d_2 \big| \ldots \big| d_n.$$

Let j *range over a complete system of representatives for* $\delta^{-1}\mathbf{Z}^n/\mathbf{Z}^n$. *Then for a fixed* r, s, *the functions*

$$\varphi(u, z, r + j, s), \quad j \in \text{representatives as above}$$

form a basis of $\mathrm{Th}(H_z, \psi_{z,r,s}, [z, \delta])$.

Proof. After multiplying by the inverse of a character, and a trivial theta function, we are reduced to proving the equivalent statement that the functions

$$\theta(u, z, r + j, s)$$

form a basis for the space $\text{Th}(z, \delta)$ of entire functions such that

$$\frac{f(u + za + \delta b)}{f(u)} = \mathbf{e}\left(-\frac{1}{2}\,{}^{t}aza - {}^{t}au\right) \quad \text{for} \quad a, b \in \mathbf{Z}^{n}.$$

But first, it is clear that these functions are linearly independent, and second the analysis of the Fourier expansion shows that the dimension of this space is $\leq d_1 \cdots d_n$ because the Fourier coefficients are determined recursively. This proves the theorem.

CHAPTER IX

Involutions and Abelian Manifolds of Quaternion Type

Certain abelian manifolds have large algebras of endomorphisms. The most common case is that of Complex Multiplication, which is treated extensively in the literature. Almost as important is the case when this algebra contains a quaternion algebra. I have therefore included this section as an example of such manifolds, which will provide easier access to their more advanced theory.

§1. Involutions

Let k be a field. By a **quaternion algebra** Q over k we mean a simple algebra with center k, of dimension 4 over k. If k has characteristic $\neq 0$, we also require that the algebra has a separable splitting field of degree 2. We are essentially concerned with quaternion algebras over number fields, so we have no intention of dwelling on the pathologies of characteristic p.

If E is a finite extension of k which splits Q, then $E \otimes Q$ (tensor product taken over k) is a semisimple algebra over E, of dimension 4 and so must be the matrix algebra $\mathbf{M}_2(E)$ of 2×2 matrices over E. The quaternion algebra Q itself is either a division algebra, or $\mathbf{M}_2(k)$. Since the algebra of 2×2 matrices over a field is simple, it admits exactly one irreducible representation of dimension 2, up to isomorphism. The trace and determinant of this representation will be denoted by **tr** and **nr** respectively, and will be called the (reduced) **trace** and **norm** of the quaternion algebra.

Let $\alpha \in Q$ but $\alpha \notin k$. Let $P_\alpha(X)$ be the minimal polynomial of α over k. Then P_α has degree >1, and hence must have degree 2. Furthermore, P_α divides the characteristic polynomial of the absolutely irreducible 2-dimensional representation of $E \otimes Q$, so P_α is equal to this characteristic polynomial. In particular, the norm and trace are those of the minimal polynomial. We let the factorization be

$$P_\alpha(X) = (X - \alpha)(X - \alpha') = X^2 - \text{tr}(\alpha)X + \text{nr}(\alpha).$$

Thus

$$\text{tr } \alpha = \alpha + \alpha' \quad \text{and} \quad \text{nr}(\alpha) = \alpha\alpha'.$$

If $\alpha \in k$, then we let $\alpha' = \alpha$.

Theorem 1.1. *The map $\alpha \mapsto \alpha'$ is an involution of Q, that is an anti-automorphism of order 2.*

Proof. The map is obviously linear. It suffices to prove the property

$$(\alpha\beta)' = \beta'\alpha'$$

when α, β are 2×2 matrices, and the trace, norm are the ordinary trace and determinant. By specialization, it suffices to prove the relation when the matrices have algebraically independent coefficients, and in particular are invertible. But then

$$\alpha' = \alpha^{-1}\det(\alpha),$$

and the relation is obvious. This proves the theorem.

We note that if F is a subfield of Q over k, then F is a quadratic extension of k, and the involution $\alpha \mapsto \alpha'$ induces the non-trivial automorphism of F if F is separable over k. Indeed, the formula $\alpha' = \alpha^{-1}\text{nr}(\alpha)$ shows that α' also lies in this subfield, and α, α' are the roots of the minimal polynomial of α over k.

The involution $\alpha \mapsto \alpha'$ will be called the **canonical involution.**

Proposition 1.2. *Every inner automorphism of Q commutes with the involution.*

Proof. Immediate from the fact that $\alpha\alpha'$ is an element of k, and so commutes with all elements of Q.

Theorem 1.3. *Let F be a separable quadratic extension of k contained in Q. Let $\varphi: F \mapsto Q$ be a k-linear embedding which is not the identity. Then there exists an inner automorphism of Q which induces φ on F. In particular, every automorphism of Q is inner.*

Proof. We shall need the remark that there is an isomorphism

$$Q \otimes Q \xrightarrow{\sim} \text{End}_k(Q_{vs})$$

where Q_{vs} denotes Q viewed as a vector space over k. The isomorphism is given as follows. An element $\Sigma \; \alpha_i \otimes \beta_i$ gives rise to an endomorphism such that

$$x \mapsto \sum \alpha_i x \beta_i'.$$

This association is a homomorphism of $Q \otimes Q$ into $\text{End}_k(Q_{vs})$. It is not identically zero, and hence is injective since $Q \otimes Q$ is simple (no two-sided ideals other than 0 and the whole algebra). It is an isomorphism since the dimensions of the domain and range are equal.

Now we may view Q as a vector space over F in two ways. First in the natural way, using the multiplication in Q. Second, by the action

$$(a, x) \mapsto \varphi(a)x \qquad \text{for } a \in F, \; x \in Q.$$

The dimension of Q over F, either way, is equal to 2. So there is an isomorphism

$$\lambda : Q \mapsto Q \text{ such that } \lambda(ax) = \varphi(a)\lambda(x),$$

and λ in particular can be viewed as an element of $\text{End}_k(Q_{vs})$. Thus by the first remark, we have

$$\lambda = \sum \alpha_i \otimes \beta_i,$$

and in particular,

$$\lambda(a) = \sum \alpha_i a \beta_i' \text{ for } a \in F.$$

The relation $\lambda(ax) = \varphi(a)\lambda(x)$ shows that

$$\sum \alpha_i a x \beta_i' = \varphi(a) \sum \alpha_i x \beta_i' \qquad \text{for all } x \in Q,$$

so

$$\sum \alpha_i a \otimes \beta_i = \sum \varphi(a)\alpha_i \otimes \beta_i.$$

Hence $\alpha_i a = \varphi(a)\alpha_i$ for all $a \in F$, and all i. Let $\alpha = \alpha_i$ for some i, $\alpha \neq 0$. We claim that α is invertible in Q. This will conclude the proof. Since $\alpha \notin F$, we have

$$Q = F + \alpha F,$$

whence

$$
\begin{aligned}
Q\alpha Q = F\alpha Q + \alpha F\alpha Q = Q\alpha F + Q\alpha\alpha F \\
\subset Q\varphi(F)\alpha + Q\alpha\varphi(F)\alpha \\
\subset Q\alpha.
\end{aligned}
$$

Hence $Q\alpha$ is a two-sided ideal $\neq 0$, whence $Q\alpha = Q$, whence α is invertible, thereby proving the theorem.

We shall now characterize all possible involutions. Let $\gamma \in Q^\times$. Define

$$\alpha^* = \alpha^*_\gamma = \gamma^{-1}\alpha'\gamma.$$

Theorem 1.4. *The map $\alpha \mapsto \alpha^*$ is an involution if and only if $\gamma^2 \in k$. Every involution is of this type, for some γ.*

Proof. Since $\alpha^{**} = \gamma^{-2}\alpha\gamma^2$, and $\alpha^{**} = \alpha$ for all $\alpha \in Q$ if and only if $\gamma^2 \in k$ (because k is the center), the first assertion is clear. Conversely, let $\alpha \mapsto \alpha^*$ be an involution. Then $\alpha \mapsto (\alpha^*)'$ is an automorphism, and so by Theorem 1.3 there exists an invertible γ such that

$$(\alpha^*)' = \gamma^{-1}\alpha\gamma \text{ for all } \alpha \in Q.$$

Since inner automorphisms commute with the involution $\alpha \mapsto \alpha'$, our theorem is proved.

The involution of Theorem 1.4 is called the **involution associated with, or defined by** γ.

§2. Special Generators

Let Q be a quaternion algebra over k, and assume for simplicity that the characteristic of k is $\neq 2$. Let F be a subfield of Q of degree 2 over k. Then

$$F = k(\beta) \text{ with some } \beta \text{ such that } \beta^2 = b \in k.$$

By the inner automorphism theorem, there exists an element $\gamma \in Q$ whose inner automorphism induces the non-trivial automorphism of F, that is $\gamma^{-1}\beta\gamma = -\beta$. Then $\gamma^{-2}\beta\gamma^2 = \beta$ so γ^2 commutes with γ and β. Since $\beta \notin k(\gamma)$ it follows that

$$Q = F + F\gamma = F + \gamma F$$

because Q is a vector space of dimension 2 over F. Therefore γ^2 lies in the center of Q, so lies in k. We have thus proved the first part of the following theorem.

Theorem 2.1. *Let F be a subfield of degree 2 over k, $F = k(\beta)$, $\beta^2 = b \in k$. Then:*
(i) There is a basis 1, β, γ, $\beta\gamma$ of Q over k such that γ is invertible, and

$$\gamma^2 = c \in k, \qquad \beta\gamma = -\gamma\beta.$$

(ii) If $z = x_0 + x_1\beta + x_2\gamma + x_3\beta\gamma$ with $x_i \in k$ then

$$\text{tr } x = 2x_0 \quad and \quad \text{nr}(z) = x_0^2 - bx_1^2 - cx_2^2 - bcx_3^2.$$

Proof. The second statement is immediate in view of the commutation rules between β and γ, and the fact that the canonical involution induces the non-trivial automorphisms of $k(\beta)$ and $k(\gamma)$ respectively, so $\beta' = -\beta$ and $\gamma' = -\gamma$.

Whenever we are in the situation of Theorem 2.1, we shall write

$$Q = (b,c)_k \quad \text{or simply} \quad (b,c), \quad \text{or} \quad (F,c).$$

Now assume that $k = \mathbf{Q}$ is the rational numbers. We say that Q is **indefinite** if $Q_{\mathbf{R}} \approx \mathbf{M}_2(\mathbf{R})$.

Theorem 2.2. *The algebra Q is indefinite if and only if Q contains a real quadratic subfield. If $Q = (b,c)$, this is the case if and only if $b > 0$ or $c > 0$.*

Proof. If Q contains a real quadratic subfield F, then $F \otimes Q \approx \mathbf{M}_2(F)$ so the splitting is clear. Conversely, suppose $Q_{\mathbf{R}} \approx \mathbf{M}_2(\mathbf{R})$. Suppose $b < 0$ and $c < 0$. Then $b = -b_1^2$ and $c = -c_1^2$ with b_1, $c_1 \in \mathbf{R}$. Put $i = \beta b_1^{-1}$ and $j = \gamma c_1^{-1}$. Then 1, i, j, ij satisfy the usual relations of the Hamilton quaternions, so $Q_{\mathbf{R}}$ must the Hamilton quaternions, which certainly do not split, a contradiction which proves the theorem.

Theorem 2.3. *Assume that $k = \mathbf{Q}$ and that Q is indefinite. Let $*$ be the involution defined by an element γ. Then $\text{tr}(\alpha\alpha^*) > 0$ for all $\alpha \in Q$, $\alpha \neq 0$ if and only if $\gamma^2 < 0$.*

Proof. Let $Q = (b,c)$ with $\gamma^2 = c$, $\beta^2 = b$. Any element α can be written in the form

$$\alpha = x + y\beta \quad \text{with } x, y \in k(\gamma).$$

A trivial computation using the commutation rule between β and γ shows that

$$
\begin{aligned}
\operatorname{tr} \alpha\alpha^* &= \operatorname{tr} (x + y\beta)\gamma^{-1}(x + y\beta)'\gamma \\
&= \operatorname{tr} (x + y\beta)(x' + \beta y') \\
&= 2(xx' + yy'b).
\end{aligned}
$$

If $\gamma^2 < 0$ then $k(\gamma)$ is an imaginary quadratic field, so xx' and $yy' \geqq 0$. If Q is indefinite, then $\beta^2 > 0$ so $\operatorname{tr} \alpha\alpha^* > 0$ for $\alpha \neq 0$. Conversely, if $\gamma^2 > 0$, pick $y = 1$. We can find x such that xx' is large negative, so $\operatorname{tr} \alpha\alpha^* < 0$. This proves the theorem.

§3. Orders

Assume that $k = \mathbf{Q}$. By a **lattice** in Q we mean a finitely generated $\mathbf{Z}$-submodule which is of rank 4. An **order** in Q is a subring which is a lattice. Let $\mathfrak{a}$ be a lattice. We define the **left order** of $\mathfrak{a}$ to be

$$
\mathfrak{o}_l(\mathfrak{a}) = \mathfrak{o} = \text{ring of all elements } \alpha \in Q \text{ such that } \alpha\mathfrak{a} \subset \mathfrak{a}.
$$

That $\mathfrak{o}$ is a ring is obvious. We must show that it is an order. Given any element $x \in Q$, and a basis $\alpha_1, \ldots, \alpha_4$ for $\mathfrak{a}$ over $\mathbf{Z}$, the elements $x\alpha_i$ can be expressed as linear combinations of this basis, with rational coefficients. Hence there exists a positive integer d such that $dx \in \mathfrak{o}$. It will now suffice to prove that $\mathfrak{o}$ has "bounded denominators", or equivalently that $\mathfrak{o}$ is finitely generated.

The bilinear map

$$
(\beta, \alpha) \mapsto \operatorname{tr}(\beta\alpha)
$$

is a non-degenerate bilinear form on Q and hence there is a dual basis $\beta_1, \ldots, \beta_4$ such that $\operatorname{tr}(\beta_j \alpha_i) = \delta_{ij}$. Let

$$
x = \sum c_j \beta_j \text{ with } c_j \in \mathbf{Q},
$$

and suppose $x \in \mathfrak{o}$. Then

$$
\operatorname{tr}(x\alpha_i) = c_i.
$$

Hence $c_i \in d^{-1}\mathbf{Z}$ where d is a common denominator for $\operatorname{tr}(\alpha_1), \ldots, \operatorname{tr}(\alpha_4)$. This proves what we wanted.

Similarly, we could define the right order $\mathfrak{o}_r(\mathfrak{a})$. We shall work with the left order.

If $\mathfrak{o}$ is the (left) order of $\mathfrak{a}$, then we also say that $\mathfrak{a}$ is an $\mathfrak{o}$-lattice, or an $\mathfrak{o}$-**ideal** if $\mathfrak{a} \subset \mathfrak{o}$. If $\mathfrak{a}$ is an $\mathfrak{o}$-lattice, then there exists a positive integer c such that $c\mathfrak{a}$ is an $\mathfrak{o}$-ideal. In fact, if $\mathfrak{a}$, $\mathfrak{b}$ are two lattices, then there exists a positive integer c such that $c\mathfrak{a} \subset \mathfrak{b}$. We also say that any two lattices are **commensurable**.

§4. Lattices and Riemann Forms on $\mathbf{C}^2$ Determined by Quaternion Algebras

We begin by some comments on a special type of lattice in $\mathbf{C}^2$.

Lemma 4.1. *Let*

$$w = \begin{pmatrix} w_1 \\ w_2 \end{pmatrix} \in \mathbf{C}^2.$$

Let $\alpha_1, \ldots, \alpha_4 \in \mathbf{M}_2(\mathbf{R})$ be linearly independent over $\mathbf{R}$. Let L be the $\mathbf{Z}$-module generated by $\alpha_1, \ldots, \alpha_4$. Then Lw is a lattice in $\mathbf{C}^2$ if and only if $w_1 w_2 \neq 0$ and $\mathrm{Im}(w_1/w_2) \neq 0$.

Proof. If w_1 or $w_2 = 0$, or if w_1, w_2 are real multiples of each other, then there is some real linear combination

$$\alpha = \sum_{i=1}^{4} c_i \alpha_i, \qquad c_i \in \mathbf{R}, \text{ not all } c_i = 0$$

such that $\alpha w = 0$, so Lw cannot be a lattice in $\mathbf{C}^2$. Conversely, if $\mathrm{Im}(w_1/w_2) \neq 0$, after multiplying the vector w by w_2^{-1} we may assume without loss of generality that

$$w = \begin{pmatrix} \tau \\ 1 \end{pmatrix}$$

with τ in the upper or lower half plane. If α is a real matrix as above, then it is clear that $\alpha w = 0$ implies $\alpha = 0$. Hence $\alpha_1 w, \ldots, \alpha_4 w$ are linearly independent over the reals, so Lw is a lattice in $\mathbf{C}^2$, thus proving the lemma.

An element w as in Lemma 4.1, with $w_1 w_2 \neq 0$ and $\mathrm{Im}(w_1/w_2) \neq 0$ will be called a **non-degenerate vector.**

The lemma will be applied to the following situation. Let Q be a quaternion algebra over the rational numbers. Let

$$\rho : Q_{\mathbf{R}} \to \mathbf{M}_2(\mathbf{R})$$

be an isomorphism. We call the pair (Q, ρ) a **quaternion type.** Let A be a complex torus of dimension 2, and let

$$\iota : Q \to \text{End}(A)_Q$$

be an embedding. We say that (A, ι) is of **type** (Q, ρ) if there exists a complex analytic isomorphism

$$\Theta : \mathbf{C}^2/\Lambda \to A$$

such that the following diagram is commutative for all $\alpha \in Q$:

$$
\begin{array}{ccc}
\mathbf{C}^2/\Lambda & \xrightarrow{\;\Theta\;} & A \\
{\scriptstyle \rho(\alpha)}\Big\downarrow & & \Big\downarrow{\scriptstyle \iota(\alpha)} \\
\mathbf{C}^2/\Lambda & \xrightarrow[\;\Theta\;]{} & A
\end{array}
$$

In fact, such an isomorphism Θ always exists. Indeed, write $A = V/\Lambda$ where V is a 2-dimensional complex space. The complex representation of $\iota(Q)$ on V is thus a 2-dimensional representation, equivalent to the representation ρ over $\mathbf{C}$ since $\mathbf{M}_2(\mathbf{C})$ is simple. Consequently, there exists a basis for V, identifying V with $\mathbf{C}^2$, such that $\iota(\alpha)$ is represented by $\rho(\alpha)$ relative to this basis.

Let $u \in \Lambda$, $u \neq 0$. The map

$$\alpha \mapsto \rho(\alpha)u, \qquad \alpha \in Q$$

is injective, and for any lattice $\mathfrak{a}$ in Q, the image $\rho(\mathfrak{a})u$ is a free $\mathbf{Z}$-module of rank 4. Furthermore, there exists a positive integer d such that $\rho(d\mathfrak{a})u \subset \Lambda$, so $\rho(d\mathfrak{a})u$ is a sublattice of Λ. Since $\rho(Q)u = Q\Lambda$ is the rational vector space generated by Λ, it follows that there exists a lattice $\mathfrak{a}$ in Q such that

$$\Lambda = \rho(\mathfrak{a})u.$$

Let $\mathfrak{o}$ be the left order of $\mathfrak{a}$. Then $\mathfrak{o}$ is also the subring of elements $\alpha \in Q$ such that $\rho(\alpha)\Lambda \subset \Lambda$. Thus

$$\mathfrak{o} = Q \cap \rho^{-1}\text{End}(A).$$

In other words, $\mathfrak{o}$ is the subring of Q corresponding to endomorphisms of A under the representation (not just endomorphisms tensored with the rationals). We can summarize our discussion in the following theorem.

Theorem 4.2. *Let $A = \mathbf{C}^2/\Lambda$ be a complex torus of dimension 2, such that (A, ι) is of type (Q, ρ). Let $u \in \Lambda$, $u \neq 0$. Then u is non-degenerate. There exists a lattice $\mathfrak{a}$ in Q such that $\Lambda = \rho(\mathfrak{a})u$. If $\mathfrak{o}$ is the order of $\mathfrak{a}$, then*

$$\rho(\mathfrak{o}) = \mathrm{End}\,(A) \cap \rho(Q).$$

Conversely, let u be a non-degenerate vector in $\mathbf{C}^2$ and let (Q, ρ) be a quaternion type. Let $\mathfrak{a}$ be a lattice in Q and let $\Lambda = \rho(\mathfrak{a})u$. Then Λ is a lattice in $\mathbf{C}^2$ and $\mathbf{C}^2/\Lambda$ is a complex torus of type (Q, ρ) as above.

The converse statement in the theorem is obvious. If we want to give all the data in the notation, we shall say that (A, ι) is of **type** $(Q, \rho, \mathfrak{a}, u)$ **with respect to** Θ. We often omit Θ from the notation, and identify A with $\mathbf{C}^2/\Lambda$ after the suitable choice of basis.

We shall now see that a torus as above is always abelian, in other words admits a Riemann form. Recall that a **Riemann form** E on V/Λ is an **R**-bilinear form on V which is skew-symmetric, non-degenerate, such that the form

$$(x, y) \mapsto E(ix, y)$$

is symmetric positive definite, and such that $E(\Lambda, \Lambda)$ is **Z**-valued.

Let E be a Riemann form on $\mathbf{C}^2/\Lambda$. We shall say that E is ρ-**admissible** if the involution determined by E leaves $\rho(Q)$ stable. In symbols, this can be written

$$\rho(Q)^* = \rho(Q).$$

We may then define an involution $\alpha \mapsto \alpha^*$ on Q such that

$$\rho(\alpha^*) = \rho(\alpha)^*.$$

By the general theory of Riemann forms, we have

$$\mathrm{tr}\,\alpha\alpha^* > 0 \qquad \text{for } \alpha \neq 0,$$

and we know from Theorem 2.3 that

$$\alpha^* = \gamma^{-1}\alpha'\gamma \text{ with some } \gamma \text{ such that } \gamma^2 < 0, \ \gamma^2 \in \mathbf{Q}.$$

Let $\{\alpha_1, \ldots, \alpha_4\}$ be a basis of Q over the rationals. Then $\{\rho(\alpha_1), \ldots, \rho(\alpha_4)\}$ is a basis of $\rho(Q_\mathbf{R})$ over **R**, and

$$\{\rho(\alpha_1)u, \ldots, \rho(\alpha_4)u\}$$

is a basis of $\rho(Q_{\mathbf{R}})u = \mathbf{C}^2$ over $\mathbf{R}$. The values of a Riemann form are determined by its values

$$E(\rho(\alpha)u, \rho(\beta)u)$$

when α, β range over such basis elements.

Theorem 4.3. *Let $\mathbf{C}^2/\Lambda$ be of type $(Q, \rho, \mathfrak{a}, u)$. Let E be a ρ-admissible Riemann form on $\mathbf{C}^2/\Lambda$. Then there exists a rational number c such that*

$$E(\rho(\alpha)u, \rho(\beta)u) = c\,\mathrm{tr}\,(\gamma\alpha\beta').$$

Conversely, given an element $\gamma \in Q$, $\gamma^2 \in \mathbf{Q}$, and $\gamma^2 < 0$, there exists an integer c such that the form determined by

$$E(\rho)(\alpha)u, \rho(\beta)u) = c\,\mathrm{tr}\,(\gamma\alpha\beta')$$

is a Riemann form which is ρ-admissible on $\mathbf{C}^2/\Lambda$.

Proof. The map $\alpha \mapsto E(\rho(1)u, \rho(\alpha)u)$ is a $\mathbf{Q}$-linear functional on Q, so there exists $\xi \in Q$ such that

$$E(\rho(1)u, \rho(\alpha)u) = \mathrm{tr}\,(\xi\alpha) \qquad \text{for all } \alpha \in Q.$$

Then

$$
\begin{aligned}
(1) \quad E(\rho(\alpha)u, \rho(\beta)u) &= E(\rho(1)u, \rho(\alpha^*\beta)u) \\
&= \mathrm{tr}\,(\xi\alpha^*\beta) = \mathrm{tr}\,(\xi\gamma^{-1}\alpha'\gamma\beta) = -\mathrm{tr}\,(\xi\gamma^{-1}\beta'\gamma\alpha)
\end{aligned}
$$

because E is anti-symmetric. Take $\beta = 1$. Then for all α we get

$$\mathrm{tr}(\gamma\xi\gamma^{-1}\alpha') = \mathrm{tr}\,(\xi\gamma^{-1}\alpha'\gamma) = -\mathrm{tr}\,(\xi\gamma^{-1}\gamma\alpha) = -\mathrm{tr}\,(\xi\alpha) = -\mathrm{tr}\,(\xi'\alpha')$$

using the fact that $\mathrm{tr}\,(\xi\alpha) = \mathrm{tr}\,(\alpha\xi)$ and $\mathrm{tr}\,(\lambda) = \mathrm{tr}\,(\lambda')$. The above relation is true for all α, and hence

$$-\gamma\xi\gamma^{-1} = \xi', \text{ so } -\gamma\xi = \xi'\gamma.$$

It follows at once that $(\xi\gamma^{-1})' = \xi\gamma^{-1}$, and therefore $\xi\gamma^{-1} \in \mathbf{Q}$, so there exists a rational number c such that

$$\xi = c\gamma.$$

From the last expression in formula (1) we find

$$E(\rho(\alpha)u, \rho(\beta)u) = -c\, \mathrm{tr}(\gamma\alpha\beta').$$

This proves the first part of the theorem.

Conversely, let $\gamma^2 \in \mathbf{Q}$, $\gamma^2 < 0$ so $\gamma' = -\gamma$. Define

$$E(\rho(\alpha)u, \rho(\beta)u) = \mathrm{tr}(\gamma\alpha\beta').$$

Then E is an anti-symmetric $\mathbf{R}$-bilinear form on $Q_{\mathbf{R}}$. We show that the form $E(iz, w)$ is symmetric. Since $\rho(Q_{\mathbf{R}}) = \mathbf{C}^2$ there exists $\eta \in Q_{\mathbf{R}}$ such that $\eta^2 = -1$,

$$iu = \rho(\eta)u, \text{ and } \eta^2 = -1, \ \eta' = \eta.$$

Since $i\rho(\alpha)u = \rho(\alpha\eta)u$, we get:

$$\begin{aligned}
E(i\rho(\alpha)u, \rho(\beta)u) = \mathrm{tr}(\gamma\alpha\eta\beta') &= \mathrm{tr}(\beta\eta\alpha'\gamma) \\
&= \mathrm{tr}(\gamma\beta\eta\alpha') = E(i\rho(\beta)u, \rho(\alpha)u).
\end{aligned}$$

This proves the symmetry.

Next, for the positive definiteness, we have

$$E(i\rho)(\alpha)u, \rho(\alpha)u) = \mathrm{tr}(\gamma\alpha\eta\alpha') = \mathrm{tr}(\alpha\eta\alpha'\gamma).$$

Let $\gamma^2 = -s^2$ with s real, and $s > 0$. By the inner automorphism theorem, there exists $\delta \in Q_{\mathbf{R}}$ such that

$$\gamma^{-1}s = \delta^{-1}\eta\delta.$$

Therefore

$$\mathrm{tr}(\alpha\eta\alpha'\gamma) = s\, \mathrm{tr}(\alpha\delta\gamma^{-1}\delta^{-1}\alpha'\gamma) = s\, \mathrm{nr}(\delta)^{-1}\, \mathrm{tr}((\alpha\delta)(\alpha\delta)^*).$$

We select c to have the sign such that $cs\, \mathrm{nr}(\delta) > 0$, to get the positive definiteness.

Finally, if α ranges over a basis $\alpha_1, \ldots, \alpha_4$ of $\mathfrak{a}$, and β ranges over such a basis also, then the finite number of elements $\mathrm{tr}(\gamma\alpha_i\beta_j')$ have bounded denominators. If we select c to be a common denominator, then the form $c \cdot \mathrm{tr}(\gamma\alpha\beta')$ is integral valued on the lattice generated by $\alpha_1, \ldots, \alpha_4$.

Since the Riemann form we have just defined is clearly ρ-admissible, this concludes the proof of the theorem.

§5. Isomorphism Classes

Let us fix a lattice $\mathfrak{a}$ with left order $\mathfrak{o}$, and a representation ρ as before. Let

$$\Lambda = \rho(\mathfrak{a})u, \quad \text{with } u = \begin{pmatrix} u_1 \\ u_2 \end{pmatrix}.$$

Suppose that u_1/u_2 is in the lower half plane. *Assume that there exists a unit* ϵ *in* $\mathfrak{o}$ *such that* $\mathrm{nr}\,(\epsilon) = -1$. Let

$$\rho(\epsilon)u = \begin{pmatrix} w_1 \\ w_2 \end{pmatrix}.$$

Then $\tau = w_1/w_2$ is in the upper half plane, and the map

$$z \mapsto w_2^{-1}\rho(\epsilon)z \qquad \text{for } z \in \mathbf{C}^2$$

gives an isomorphism

$$\mathbf{C}^2/\Lambda \xrightarrow{\ \sim\ } \mathbf{C}^2/\Lambda', \qquad \text{where } \Lambda' = \rho(\mathfrak{a})\begin{pmatrix} \tau \\ 1 \end{pmatrix}.$$

Thus we see that to study isomorphism classes of abelian manifolds admitting quaternion multiplication, we may limit ourselves to representatives whose lattices are of the form

$$\Lambda_\mathfrak{a}(\tau) = \rho(\mathfrak{a})\begin{pmatrix} \tau \\ 1 \end{pmatrix}, \qquad \text{with } \tau \in \mathfrak{H}, \text{ (The upper half plane)}.$$

We now derive a necessary and sufficient condition that the abelian manifolds be isomorphic.

Assume that $\mathfrak{a} = \mathfrak{o}$. Let us use the notation

$$A(\tau) = \mathbf{C}^2/\Lambda(\tau), \quad \text{where } \Lambda(\tau) = \rho(\mathfrak{o})\begin{pmatrix} \tau \\ 1 \end{pmatrix}.$$

Consider a homomorphism

$$h : (A(\tau_1),\rho) \to (A(\tau_2),\rho)$$

which commutes with the representation ρ. Such h is represented by a complex matrix M on $\mathbf{C}^2$ which commutes with $\rho(\alpha)$ for all $\alpha \in Q$, and therefore M is a scalar,

(1) $M = gI_2$ with $g \in \mathbf{C}$.

We suppose $h \neq 0$. There exists an element $\lambda \in \mathfrak{o}$ such that

(2) $$M\binom{\tau_1}{1} = \rho(\lambda)\binom{\tau_2}{1}$$

because $M\Lambda(\tau_1) \subset \Lambda(\tau_2)$. We let $GL_2(\mathbf{R})$ operate on complex numbers with non-zero imaginary part as usual. From (1) and (2), we conclude that

(3) $$\mathrm{nr}(\lambda) = \det \rho(\Lambda) > 0$$

because τ_1 and τ_2 have imaginary parts with the same sign.

We then apply this discussion to isomorphisms.

Theorem 5.1. *Assume that* $\mathfrak{a} = \mathfrak{o}$. *Then* $(A(\tau_1),\rho)$ *and* $(A(\tau_2),\rho)$ *are isomorphic if and only if there exists a unit* ϵ *in* $\mathfrak{o}$ *with* $\mathrm{nr}(\epsilon) = 1$ *such that* $\rho(\epsilon)(\tau_1) = \tau_2$. *Isomorphisms are then given by the matrix operations* $\rho(\epsilon)$ *for such elements* ϵ.

Proof. In the discussion preceding the theorem, h is an isomorphism if and only if $M\Lambda(\tau_1) = \Lambda(\tau_2)$, or equivalently

$$\rho(\mathfrak{o})\rho(\lambda) = \rho(\mathfrak{o}).$$

This is equivalent with $\mathfrak{o}\lambda = \mathfrak{o}$, that is λ is a unit in $\mathfrak{o}$, and we already know by (3) that $\mathrm{nr}(\lambda) = 1$. Conversely, given such a unit λ, let

$$\rho(\lambda) = \begin{pmatrix} a & b \\ c & d \end{pmatrix},$$

and let $g = c\tau_2 + d$. Then it follows at once that

$$\rho(\lambda)\binom{\tau_1}{1} = g\binom{\tau_1}{1}.$$

Since λ is a unit in $\mathfrak{o}$, we have $g\Lambda(\tau_1) = \Lambda(\tau_2)$. Hence λ induces an isomorphism as stated in the theorem.

CHAPTER X

Theta Functions and Divisors

Let M be a complex manifold. In the sequel, M will either be $\mathbf{C}^n$ or $\mathbf{C}^n/D$, where D is a lattice (discrete subgroup of real dimension $2n$). Let $\{U_i\}$ be an open covering of M, and let φ_i be a meromorphic function on U_i. If for each pair of indices (i, j) the function φ_i/φ_j is holomorphic and invertible on $U_i \cap U_j$, then we shall say that the family $\{(U_i, \varphi_i)\}$ represents a divisor on M. If this is the case, and (U, φ) is a pair consisting of an open set U and a meromorphic function φ on U, then we say that (U, φ) is **compatible** with the family $\{(U_i, \varphi_i)\}$ if φ_i/φ is holomorphic invertible on $U \cap U_i$. If this is the case, then the pair (U, φ) can be adjoined to our family, and again represents a divisor. Two families $\{(U_i, \varphi_i)\}$ and $\{(V_k, \psi_k)\}$ are said to be **equivalent** if each pair (V_k, ψ_k) is compatible with the first family. An equivalence class of families as above is called a **divisor** on M. Each pair (U, φ) compatible with the families representing the divisor is also said to represent the divisor on the open set U.

If $\{(U_i, \varphi_i)\}$ and $\{(V_k, \psi_k)\}$ represent divisors, then it is clear that

$$\{(U_i \cap V_k, \varphi_i\psi_k)\}$$

also represents a divisor, called the **sum**.

For simplicity of language, we sometimes say that the family $\{(U_i, \varphi_i)\}$ is itself a divisor, say X, and write $X = \{(U_i, \varphi_i)\}$. We say that X is **positive** if it has a representative family in which all the functions φ_i are holomorphic.

If φ is a meromorphic function on M, then (M, φ) represents a divisor, and we say in this case that φ **represents this divisor globally**.

The result of this chapter and its proof will be independent of everything that precedes. We need only know the definition of a theta function: let V be a complex vector space of dimension n, and let D be a lattice in V. For this chapter we assume our theta functions to be entire, and so a theta function on V with respect to D will be an entire function F, not identically zero, satisfying the condition

$$F(z + u) = F(z)e^{2\pi i[L(z,u)+J(u)]},$$

where L is **C**-linear in z, and the above equation holds for all $u \in D$. We shall also identify V with $\mathbf{C}^n$, with respect to a fixed basis, and then the exponential term obviously can be rewritten for each u in the form

$$L(z, u) + J(u) = \sum c_\alpha z_\alpha + b,$$

where c_α and b are complex numbers, depending on u. Thus the exponent is a polynomial in z of degree 1, with coefficients depending on u.

We shall prove that given a divisor X on $\mathbf{C}^n/D$, there exists a theta function F representing this divisor on $\mathbf{C}^n$. If the reader knows the content of Chapter VI, he will then realize that there is a bijection between divisors on the torus and normalized theta functions (up to constant factors), and this bijection is homomorphic, i.e., to the sum of two divisors corresponds the product of their normalized theta functions.

Furthermore, two (entire) theta functions have the same divisor if and only if they are equivalent (i.e., differ by a trivial theta function). Finally, since to each theta function we can associate a Riemann form, we see that we can associate a Riemann form with a divisor, uniquely, and that this association is additive.

We now turn to the existence theorem, whose proof is self-contained, that is, makes no use of the linear theory developed in the previous chapters.

§1. Positive Divisors

Theorem 1.1. *Let $\overline{X}$ be a positive divisor on $\mathbf{C}^n/D$, and let X be its inverse image on $\mathbf{C}^n$. Then there exists an entire theta function F representing this divisor on $\mathbf{C}^n$.*

Proof. The proof will be carried out by juggling with differential forms, and reproving ad hoc some results valid on Kähler manifolds. Everything becomes much simpler because we work on the torus and $\mathbf{C}^n$.

Lemma 1. *Let M be a C^∞ manifold, and $\{U_i\}$ a locally finite open covering. For each pair (i, j) such that $U_i \cap U_j$ is not empty, suppose given a differential form ω_{ij} of degree p, satisfying*

$$\omega_{ij} - \omega_{ik} + \omega_{jk} = 0$$

in $U_i \cap U_j \cap U_k$ whenever this intersection is not empty. Then there exist differential forms ω_i on U_i such that

$$\omega_{ij} = \omega_i - \omega_j$$

on $U_i \cap U_j$, whenever this intersection is not empty.

Proof. Let $\{g_i\}$ be a partition of unity subordinated to the given covering. We let

$$\omega_i = \sum_j g_j \omega_{ij},$$

with the obvious convention that the expression on the right is equal to 0 wherever it is not defined. Using the cocycle equation, and its obvious consequences that

$$\omega_{ii} = 0, \qquad \omega_{ij} = -\omega_{ji},$$

we get our lemma.

The next lemma again considers only C^∞ forms. Let D be a lattice in $\mathbf{R}^m$, and $T = \mathbf{R}^m/D$ the torus. Let $x_1, \ldots, x_m$ be the real coordinates of $\mathbf{R}^m$, and write a p-form as

$$\sum f_{i_1 \cdots i_p} \, dx_{i_1} \wedge \cdots \wedge dx_{i_p},$$

taking the sum over the indices $i_1 < \cdots < i_p$. Let $I(\omega)$ be the form with constant coefficients obtained by replacing each function $f_{(i)}$ by its integral

$$I(f_{(i)}) = \int_T f_{(i)}(x) \, dx.$$

Since $f_{(i)}$ can be viewed as a periodic function on $\mathbf{R}^m$, we can view the integral as a multiple integral after a change of variables if necessary.

Define

$$\frac{\partial \omega}{\partial x_j} = \sum \frac{\partial}{\partial x_j} f_{(i)} \, dx_{i_1} \wedge \cdots \wedge dx_{i_p}.$$

In other words, define the partial derivative of the form to be obtained by applying it to the functions $f_{(i)}$. Similarly, if Δ is a differential operator, we denote by $\Delta\omega$ the form obtained from ω, replacing all functions $f_{(i)}$ by $\Delta f_{(i)}$.

Integrating by parts, we see at once that

$$I\left(\frac{\partial \omega}{\partial x_j}\right) = 0.$$

Lemma 2. *Let (a_{ij}) be a real symmetric positive definite matrix. Let*

$$\Delta = \sum a_{ij} \frac{\partial^2}{\partial x_i \, \partial x_j}.$$

Let ω be a p-form on the torus. There exists a p-form ψ on the torus such that $\Delta\psi = \omega$ if and only if $I(\omega) = 0$. If $\Delta\psi = 0$, then ψ has constant coefficients.

Proof. Since our operators I and Δ actually operate on functions, we can just deal with functions f on the torus, viewed as periodic functions on $\mathbf{R}^m$. Since these functions are assumed to be C^∞, they have Fourier expansions which converge rapidly to 0, as one sees by the usual integration by parts. Say

$$f(x) = \sum c_\nu e^{2\pi i \nu \cdot x},$$

where the sum is taken over $\nu = (\nu_1, \ldots, \nu_m)$, and $\nu \cdot x$ is the dot product. Note that

$$\Delta(e^{2\pi i \nu \cdot x}) = (2\pi i)^2 Q(\nu) e^{2\pi i \nu \cdot x},$$

where $Q(\nu) = \sum a_{jk} \nu_j \nu_k$ is the value of the quadratic form at ν. If $I(f) = 0$, then the constant term in the Fourier expansion is equal to 0, and we can then solve trivially term by term for the Fourier coefficients of a function g such that $\Delta g = f$. The converse is trivial (integration by parts). Also, if $\Delta g = 0$, then we see at once from the way Δ operates on $e^{2\pi i \nu \cdot x}$ that g must be constant. This proves our lemma.

We now take $\mathbf{R}^m = \mathbf{C}^n$ ($m = 2n$). We shall use the usual coordinates z_α and $\bar{z}_\alpha$, where

$$z_\alpha = x_\alpha + iy_\alpha, \qquad \bar{z}_\alpha = x_\alpha - iy_\alpha.$$

We define

$$\frac{\partial}{\partial z_\alpha} = \frac{1}{2}\left(\frac{\partial}{\partial x_\alpha} + \frac{1}{i}\frac{\partial}{\partial y_\alpha}\right) \qquad \text{and} \qquad \frac{\partial}{\partial \bar{z}_\alpha} = \frac{1}{2}\left(\frac{\partial}{\partial x_\alpha} - \frac{1}{i}\frac{\partial}{\partial y_\alpha}\right),$$

because we can solve for x_α and y_α in terms of z_α, $\bar{z}_\alpha$, and then

$$\frac{\partial x_\alpha}{\partial z_\alpha} = \frac{1}{2}, \qquad \frac{\partial y_\alpha}{\partial z_\alpha} = \frac{1}{2i},$$

$$\frac{\partial x_\alpha}{\partial \bar{z}_\alpha} = \frac{1}{2}, \qquad \frac{\partial y_\alpha}{\partial \bar{z}_\alpha} = -\frac{1}{2i}.$$

We express the differential forms in terms of dz_α and $d\bar{z}_\alpha$, and take the Laplacian to be

$$\Delta = \sum \frac{\partial^2}{\partial z_\alpha\, \partial \bar{z}_\alpha}.$$

If a differential form is written as

$$\omega = f_{(\alpha,\beta)}\, dz_{\alpha_1} \wedge \cdots \wedge dz_{\alpha_p} \wedge d\bar{z}_{\beta_1} \wedge \cdots \wedge d\bar{z}_{\beta_q},$$

then we say that it is of **type (p, q)**. Its exterior derivative is given by

$$d\omega = \sum_j \frac{\partial f_{(\alpha,\beta)}}{\partial z_j}\, dz_j \wedge dz_{(\alpha)} \wedge d\bar{z}_{(\beta)} + \sum_j \frac{\partial f_{(\alpha,\beta)}}{\partial \bar{z}_j}\, d\bar{z}_j \wedge dz_{(\alpha)} \wedge d\bar{z}_{(\beta)}.$$

In other words, the same formalism prevails as with the real coordinates. A C^∞ function f on U is holomorphic if and only if for all α,

$$\frac{\partial f}{\partial \bar{z}_\alpha} = 0 \qquad \text{(Cauchy-Riemann equations)}.$$

For one variable this is standard. For several variables, the Cauchy-Riemann equations in each variable show that f is holomorphic in each variable separately. By repeated use of the Cauchy formula in one variable, one then gets a power series expansion of f in all variables, because f is continuous, whence one sees that f is holomorphic in several variables.

We now suppose given a (complex) positive divisor on the torus $\mathbf{C}^n/D$, represented by, say, a finite covering $\{(U_i, \varphi_i)\}$. We also assume that U_i is the image of a ball U_{i0} in $\mathbf{C}^n$ under the canonical homomorphism $\mathbf{C}^n \to \mathbf{C}^n/D$. Then φ_i, viewed as a periodic function on $\mathbf{C}^n$, lifts in particular to a function φ_{i0} on U_{i0}. For any lattice point $l \in D$, we let U_{il} be the translate of U_{i0} by l. Then φ_i lifts to φ_{il} on U_i. Note that the balls $\{U_{il}\}_{i,l}$ form an open covering of $\mathbf{C}^n$.

Using Lemma 1, we can write

$$\zeta_{ij} = d \log \varphi_i/\varphi_j = \zeta_i - \zeta_j,$$

where ζ_i is a C^∞, 1-form on U_i. Since ζ_{ij} is of type $(1, 0)$, it follows that for all indices α, we have

$$\frac{\partial \zeta_i}{\partial \bar{z}_\alpha} = \frac{\partial \zeta_j}{\partial \bar{z}_\alpha} \quad \text{on} \quad U_i \cap U_j.$$

Hence there is a 1-form η_α on the torus, equal to $\partial \zeta_i / \partial \bar{z}_\alpha$ on each U_i. Let

$$\gamma = \sum_\alpha \frac{\partial \eta_\alpha}{\partial z_\alpha}.$$

Then $\gamma = \Delta \zeta_i$ on U_i. But $I(\gamma) = 0$. Hence by Lemma 2 there exists a 2-form ζ such that $\gamma = \Delta \zeta$. Let

$$\zeta_i' = \zeta_i - \zeta.$$

Then

$$\Delta \zeta_i' = 0 \quad \text{and} \quad \zeta_{ij} = \zeta_i' - \zeta_j'.$$

But $\zeta_{ij} = d \log \varphi_i / \varphi_j$ is of type $(1, 0)$. Hence we need only the $(1, 0)$ part of ζ_i' and ζ_j' for this relation. We let ζ_i'' be the $(1, 0)$ part of ζ_i'. Then

$$d \log \varphi_i / \varphi_j = \zeta_i'' - \zeta_j''$$
$$\Delta \zeta_i'' = 0 \quad \text{on} \quad U_i.$$

We have $d \zeta_i'' = d \zeta_j''$ on $U_i \cap U_j$ because $d^2 = 0$. Hence there exists a 2-form ω on the torus such that $\omega | U_i = d \zeta_i''$. Since

$$d\Delta = \Delta d,$$

it follows that $\Delta \omega = 0$, and hence ω has constant coefficients by Lemma 2. Since ζ_i'' is of type $(1, 0)$, we can write

$$\omega = \sum a_{\alpha\beta} \, dz_\alpha \wedge dz_\beta + \sum b_{\alpha\beta} \, d\bar{z}_\alpha \wedge dz_\beta.$$

Let

$$\psi = \sum a_{\alpha\beta} z_\alpha \, dz_\beta + \sum b_{\alpha\beta} \bar{z}_\alpha \, dz_\beta \quad \text{on} \quad \mathbf{C}^n.$$

Then $d\psi = \omega$.

Let $U_{il} = U_{i0} + l$, where $l \in D$ is a lattice point. Then

$$d(\zeta_i'' - \psi) = 0 \quad \text{on} \quad U_{il}.$$

By Poincaré's lemma, there exists a C^∞ function f_{il} on U_{il} such that

$$df_{il} = \zeta_i'' - \psi.$$

Since this is of type $(1, 0)$, we conclude that f_{il} is holomorphic by the Cauchy-Riemann equations. But on $U_{il} \cap U_{jl'}$ we have

$$df_{il} - df_{jl'} = \zeta_i'' - \zeta_j'' = d \log \varphi_i / \varphi_j.$$

Hence

$$\varphi_i e^{-f_{il}} \quad \text{and} \quad \varphi_j e^{-f_{jl'}}$$

differ by a constant multiple on $U_{il} \cap U_{jl'}$. Consequently, starting with say $\varphi_1 e^{-f_{10}}$, we can continue analytically to a function F on $\mathbb{C}^n$ which differs from $\varphi_i e^{-f_{il}}$ by a constant multiple on U_i.

We now contend that F is our desired theta function, namely

$$F(z + l) = F(z) e^{2\pi i [L(z, l) + J(l)]}.$$

Say $z \in U_{i0}$. Then

$$\frac{F(z + l)}{F(z)} = c \cdot \frac{\varphi_i(z + l)}{\varphi_i(z)} \frac{e^{-f_{il}(z + l)}}{e^{-f_{i0}(z)}},$$

and since φ_i is periodic, we get

$$
\begin{aligned}
d \log F(z + l)/F(z) &= -df_{il}(z + l) + df_{i0}(z) \\
&= \zeta_i''(z + l) - \psi(z + l) - \zeta_i''(z) + \psi(z) \\
&= \sum a_{\alpha\beta}(z_\alpha + l) \, dz_\beta + \sum b_{\alpha\beta}(\bar{z}_\alpha + \bar{l}) \, dz_\beta \\
&\quad - \left(\sum a_{\alpha\beta} z_\alpha \, dz_\beta + \sum b_{\alpha\beta} \bar{z}_\alpha \, dz_\beta \right) \\
&= \sum c_\beta(l) \, dz_\beta.
\end{aligned}
$$

This is a 1-form, with coefficients depending only on l. Integrating with respect to z gives what we wanted, and proves the theorem.

§2. Arbitrary Divisors

Let $\{(U_i, \varphi_i)\}$ represent an arbitrary divisor, not necessarily positive. The open sets U_i may be taken arbitrarily small. In this section, we do not give complete proofs. We assume the fact that the ring of convergent power series in the neighborhood of a point is a unique factorization domain. This means that if U is a sufficiently small open set around a given point, a meromorphic functions φ on U, being the quotient of two holomorphic functions on U, has an expression

$$\varphi = g/h,$$

where g, h are relatively prime, that is are not divisible by the same irreducible element. Thus we can write each $\varphi_i = g_i/h_i$. Then on $U_i \cap U_j$ we know that

$$g_i g_j^{-1}/h_i h_j^{-1}$$

is invertible holomorphic. Since g_i, h_i are relatively prime, it follows that $g_i g_j^{-1}$ is itself a unit on that intersection. Hence $\{(U_i, g_i)\}$ represents a positive divisor, as does $\{(U_i, h_i)\}$. Thus from the unique factorization we can decompose a divisor as a difference of two positive ones. In that way, the quotient of the theta functions associated with these positive divisors will represent the given divisor globally.

§3. Existence of a Riemann Form on an Abelian Variety

We wish to indicate a proof that if there is a projective embedding

$$\theta : V/D \rightarrow A_{\mathbf{C}}$$

of a torus onto the complex points of a projective variety A, then (V, D) admits a non-degenerate Riemann form. We assume that the reader is now acquainted with the terminology of algebraic geometry and abelian varieties.

Let X be a hyperplane section of A. In the neighborhood of each point, X can be defined by a local equation $\varphi = 0$, so X can also be viewed as a divisor in the sense we have used previously. Then $\theta^{-1}(X)$ is a divisor on V, and has an associated theta function θ_0, which is entire since X is a positive divisor. Let H be the associated hermitian form. The meromorphic functions giving the projective embedding $\{f_j\}$ can be written in the form $f_j = \theta_j/\theta_0$ where

$$\theta_j \in \mathscr{L}(\theta_0)$$

in the sense of Chapter VI, §4. If V_H is the kernel of H, then by Theorem 2.1 of Chapter VI, we know that θ_j factor through V/V_H. If $V_H \neq 0$, then V/V_H has dimension strictly less than n, and this contradicts Corollary 1 of Theorem 4.1 of Chapter VI, because n of the functions among the f_j are algebraically independent. This concludes the proof.

Bibliography

[Fa 1]   D. K. FADDEEV, *On the divisor class groups of some algebraic curves*, Dokl. Tom 136 pp. 296–298 = Sov. Math. Vol. 2 (1961) pp. 67–69.

[Fa 2]   D. K. FADDEEV, *Invariants of Divisor classes for the curves* $x^k(1 - x) = y^l$ *in l-adic cyclotomic fields*, Trudy Math. Inst. Steklov 64 (1961) pp. 284–293.

[Ko-R]   N. KOBLITZ and D. ROHRLICH, *Simple factors in the Jacobian of a Fermat curve*, Can. J. Math. XXX No. 6 (1978) pp. 1183–1205.

[La 1]   S. LANG, *Abelian varieties*, Interscience, 1958.

[La 2]   S. LANG, *Algebraic number theory*, Addison-Wesley, 1970.

[La 3]   S. LANG, *Complex multiplication*, Springer Verlag, 1983.

[Mu]   D. MUMFORD, *Abelian varieties*, Oxford University Press, 1970.

[Ro 1]   D. ROHRLICH, *The periods of the Fermat curve*, Appendix to B. Gross, Invent. Math. 45 (1978) pp. 193–211.

[Ro 2]   D. ROHRLICH, *Points at infinity on the Fermat curves*, Invent. Math. 39 (1977) pp. 95–127.

[Sh]   G. SHIMURA, *On the derivatives of theta functions and modular forms*, Duke Math. J. (1977) pp. 365–387.

[Sh-T]   G. SHIMURA and Y. TANIYAMA, *Complex multiplication of abelian varieties*, Mathematical Society of Japan, Publication No. 6, 1961.

[Si 1]   C. L. SIEGEL, *Lectures on several complex variables*, Institute for Advanced Studies, Princeton, reprinted 1962.

[Si 2]   C. L. SIEGEL, *Topics in complex function theory*, Interscience, 1970–1971.

[We 1]   A. WEIL, *Théorème fondamentaux de la théorie des fonctions theta*, Seminaire Bourbaki, May 1949.

[We 2]   A. WEIL, *Varietés Kählériennes*, Hermann, Paris, 1958.

Index